L. G.
1.60 net   32/— net

MATHEMATICAL PHYSICS SERIES
Editor: G. Stephenson, B.Sc., Ph.D., D.I.C.
*Department of Mathematics, Imperial College, London*

# AN INTRODUCTION TO THE THEORY OF ELECTROMAGNETIC WAVES

# An Introduction to
# THE THEORY OF
# ELECTROMAGNETIC WAVES


A. C. HEWSON

M.A., D.PHIL.

*Lecturer in Mathematics*
*Imperial College, London*

LONGMAN

LONGMAN GROUP LIMITED
London

*Associated companies, branches and representatives
throughout the world*

© *Longman Group Ltd* 1970



*First published* 1970

ISBN 0 582 46320 3

*Printed in Great Britain by
T. and A. Constable Ltd, Hopetoun Street, Edinburgh*

# CONTENTS

# Contents

## 5. Sources of Electromagnetic Waves



## 6. Scattering and Diffraction

# PREFACE

This book is based on a third-year undergraduate course in the Mathematics Department at Imperial College. The subject-matter has been broadened and the treatment modified to be suitable for third-year courses on electromagnetic waves both for physicists and electrical engineers. The background knowledge required on the reader's part has been kept to a minimum, though it is assumed that he is conversant with the mathematical methods of vector analysis, matrices, Fourier series, and the method of separation of variables for partial differential equations, and that he has attended a first- or second-year course on electricity and magnetism.

The aim of the presentation has been to introduce the physical ideas and assumptions rather carefully, and to approach the theory in a systematic way so that it can be viewed in a wider mathematical context. Starting from a preliminary discussion of electromagnetic fields, Maxwell's equations and boundary conditions, various types of wave solutions in unbounded media are examined, including cases involving dispersion, absorption, and anisotropy. The introduction of boundaries leads the discussion on to the topics of reflection, refraction, and to problems of wave propagation along wave guides. Potentials are introduced to deal with wave sources, and the concept of a retarded potential is used in the calculation of radiation from various forms of charge and current distributions. The final chapter is concerned with problems in the scattering and diffraction of plane waves. In the course of the development the way has been prepared for such advanced techniques as the use of generalized functions, Green's functions, stationary phase methods, and asymptotic expansions, which are normally only introduced at postgraduate level.

The arguments are self-contained in nearly all cases, and the reader is expected to follow the equations line by line and to carry out any intermediate manipulations that are indicated. This is essential to gain a working knowledge of the subject. Problems are given at the end of each chapter and should not be omitted in the course of reading.

I am very conscious of the fact that many topics that come under the general heading of electromagnetic waves have not been included;

I can only hope that in being brief and selective, the reader will be left with enough enthusiasm to pursue these topics for himself from the further reading given in the Bibliography.

I am very grateful to Dr F. G. Leppington for reading through a preliminary draft of the manuscript, and for his many constructive suggestions.

Imperial College, London
1969.

A. C. H.

**CHAPTER 1**

# Fundamentals of Electromagnetism

## 1.1 Maxwell's equations

Electromagnetic waves are classified into several different types but they are all part of a single continuous spectrum. There are radio waves which have wavelengths of the order of $10^2$ metres, microwaves with wavelengths of the order of $10^{-2}$ m, infra-red $10^{-5}$ m, optical waves $10^{-7}$ m, X-rays $10^{-10}$ m, and $\gamma$-rays whose lower wavelength limit is not known. These waves are such universal phenomena that it is remarkable that their existence was first predicted on theoretical grounds by James Clerk Maxwell (1831–79), when he deduced a wave equation from his formulation and generalization of the laws of electromagnetism. The velocity of the waves described by his equation depended upon certain electrical constants, and their measurement indicated that this velocity was the same as the velocity of light, $3 \times 10^8$ m/sec. This naturally led Maxwell to the conjecture that light was an electromagnetic waveform. Previous to his work, there was some hint of this as a possibility in the experimental discovery by Michael Faraday (1791–1867) that the plane of polarization of light was rotated by the presence of a magnetic field. Direct evidence in support of Maxwell's theory, however, came only in 1888 in the experiments of Heinrich Hertz (1857–94), twenty-seven years after Maxwell's original prediction.

Maxwell's theory was based on the equation

$$\operatorname{div} \boldsymbol{D} = \rho \qquad (1.1), \qquad \operatorname{div} \boldsymbol{B} = 0 \qquad (1.2),$$

$$\operatorname{curl} \boldsymbol{H} = \frac{\partial \boldsymbol{D}}{\partial t} + \boldsymbol{J} \quad (1.3), \qquad \operatorname{curl} \boldsymbol{E} = -\frac{\partial \boldsymbol{B}}{\partial t}, \quad (1.4)$$

where the vector operators, div and curl, are defined in the usual way (see Appendix B). These equations which relate the electric fields $\boldsymbol{E}$ and $\boldsymbol{D}$, the magnetic fields $\boldsymbol{B}$ and $\boldsymbol{H}$, and the charge and current densities $\rho$ and $\boldsymbol{J}$, have become the basis for the quantitative description of all types of electromagnetic waves, except those of extremely

1

small wavelength. They have survived the revolutions in physical ideas that came in the twentieth century with the theory of relativity, atomic theory and quantum mechanics. It was shown by Einstein (1879–1955) that the assumption of the constancy of the velocity of light in all inertial frames, which is basic to the theory of relativity, is consistent with Maxwell's equations. The equations are covariant under the Lorentz transformation of special relativity. Atomic theory and quantum mechanics have helped to delineate more clearly the sphere of applicability of the equations which, loosely speaking, is to phenomena on a macroscopic scale. Problems such as the emission and absorption of light or the scattering of X-rays, which are processes at an atomic level, have to be treated using a quantum mechanical model of the atom and quantized fields.

We shall assume that the reader is familiar with the historical development of ideas leading to Maxwell's equations, so that we can take (1.1) to (1.4) as our starting-point. However, at the risk of criticism for being too cursory or for starting at too elementary a level, we shall review some of the evidence on which these equations are based in order to develop some intuitive feel for their physical meaning. By examining the historical evidence in the light of our present atomic picture of matter we hope to gain insight into the assumptions which enable such relatively simple equations to be applied to such complex physical systems as metals and insulators.

## 1.2 Charge

The first of Maxwell's equations (1.1) is a generalization of the inverse square law

$$F_{12} = C_{12} \frac{r_1 - r_2}{|r_1 - r_2|^3}. \tag{1.5}$$

This was based on the experiments of Coulomb (1786–1806) in which the force $F_{12}$ between two small charged pith balls with position vectors $r_1$ and $r_2$ was measured. It was found that the constant of proportionality $C_{12}$, depended upon the electrical state of the two balls and could be factorized as $C_{12} = K_0 q_1 q_2$, $q_1$ specifying the state of the first ball and termed its charge, and $q_2$ the corresponding quantity for the second ball. Units of charge were originally defined by choosing $K_0$ to be unity, but this led to a system of units of an inconvenient magnitude for practical purposes. To bring (1.5) in line with the current trend to S.I. units (Système International d'Unités), we define $K_0$ as $1/4\pi\varepsilon_0$, where $\varepsilon_0$ has the value $8\cdot854 \times 10^{-12}$ farads/metre and is known as the *permittivity* or *dielectric constant* of free space. The criteria for a reasonable system of units are rather involved and we postpone a discussion of this topic to Appendix A.

The direction of the force was understood only by making the assumption that two types of charge exist, positive and negative, so that like charges repel and unlike charges attract. Charge, it was also discovered, could be combined arithmetically, the total charge on a body formed by bringing together charges $q_1$, $q_2$ ... $q_n$ being $q_1 + q_2$ ... $+ q_n$, with the appropriate signs. This naturally led to the indestructibility of charge as a fundamental principle of Nature.

Our present understanding of charge is that it is an intrinsic property of elementary particles. The particles that are the primary constituents of matter, the electron, the proton and the neutron have charges $-e$, $+e$ and zero respectively, where $e = 1\cdot6 \times 10^{-19}$ Coulombs. The electron has a mass $(m_e)$ of $9\cdot1 \times 10^{-31}$ kg, the proton is much heavier with a mass $1836.1\ m_e$, and the neutron has a mass $1838.7\ m_e$ almost the same as that of the proton.

An atom is electrically neutral and is composed of a positively charged nucleus surrounded by electrons in motion. The nucleus has dimensions of the order of $10^{-15}$ m and carries a total charge $Ze$ as it is composed of $Z$ protons plus about the same number, or rather more, neutrons, where $Z$ is the atomic number. The $Z$ electrons, which compensate the charge of the nucleus, are in certain distinct energy levels or atomic 'shells' about the nucleus, such that the total atom has an effective radius of $10^{-10}$ m. In any material the presence of the charge in the constituent atoms becomes apparent on a *macroscopic scale* only when the atomic balance of a large number of atoms is disturbed in some way.

### 1.3 The electric field and electric displacement

The electric field or electric intensity $E$ was first introduced to simplify the analysis of system of charges. Equation (1.5) can be separated into two stages

$$F_{12} = q_1 E_2(r_1), \tag{1.6}$$

and

$$E_2(r_1) = \frac{q_2}{4\pi\varepsilon_0} \frac{r_1 - r_2}{|r_1 - r_2|^3}. \tag{1.7}$$

The force on $q_1$ is regarded as being due to an electric field $E_2$ at $r_1$

that is generated by the charge $q_2$. As the field is linearly proportional to the charge, the field due to a system of charges $q_1, \ldots, q_n$ is the vector sum of the individual fields

$$E(r) = \frac{1}{4\pi\varepsilon_0} \sum_{s=1}^{n} q_s \frac{r - r_s}{|r - r_s|^3}. \tag{1.8}$$

The introduction of a field appears from this point of view to be merely a mathematical artifice to facilitate the description of action at a distance. However, it does relate to the ideas of Faraday that a charge generates 'lines of force' and that space is in some state of excitation. Faraday's ideas inspired Maxwell in formulating his equations and in his interpretation of them through mechanical analogies. They led to the concept of the ether, an all pervading medium which was thought to be in different states of stress due to the electric fields. Though this concept was found to be untenable and abandoned with the general acceptance of the theory of relativity, we shall find that when we come to consider electromagnetic waves, which are fields propagated with a finite velocity, we are almost led to interpret the fields as having some form of physical existence.

Equation (1.8) can be used to prove an equivalent statement of the inverse square law due to Gauss (1777–1855), that the total flux $\int E \cdot dS$ over a simple closed surface $S$ bounding a volume $V$, is equal to the total charge within $V$ divided by $\varepsilon_0$

$$\int_S E \cdot dS = \sum_{s=1}^{n} \frac{q_s}{\varepsilon_0}. \tag{1.9}$$

Charge, we know, is always in discrete units of $e$ but it is reasonable to assume when dealing with phenomena on a macroscopic scale that within a continuous medium charge can be described by a continuous distribution function or charge density $\rho(r)$. If this assumption is made, (1.9) can be written in the form

$$\int_S E \cdot dS = \frac{1}{\varepsilon_0} \int_V \rho\,d\tau. \tag{1.10}$$

The divergence theorem can be used to express the surface integral as a volume integral of div $E$. The resulting equation is true for any volume $V$, so it can be replaced by the point relation

$$\text{div } E = \rho/\varepsilon_0. \tag{1.11}$$

This is Poisson's (1781–1840) generalization of the inverse square law.

Most substances can be classified into one of two types, according to their behaviour in the presence of an electric field, being either insulators or conductors. In insulators, such as glass, the electrons in the constituent atoms (or molecules) can be displaced only over atomic distances by an applied field and remain bound to their parent atom (or molecule). Though the atoms remain neutral, there is change of charge distribution, the centre of gravity of the electron cloud does not coincide with that of the nucleus; consequently, the atom has an induced dipole moment and associated electric fields. The response of an insulator to an applied field, its polarization, is measured by the induced dipole moment $r\rho(r)$ per unit volume. This is the moment of the charges averaged over a volume which is small on a macroscopic scale but large as compared with the atomic dimensions. It can be shown that the charge carried out of a volume $V$ enclosed by a surface $S$ when a polarization $P$ is induced is $-P \cdot dS$. Equation (1.10) can be generalized to apply to an insulator

$$\int_S \left( E + \frac{P}{\varepsilon_0} \right) \cdot dS = 1/\varepsilon_0 \int_V \rho\, d\tau. \tag{1.12}$$

The polarization term allows for the charge displaced from the volume due to the applied field $E$. Using the divergence theorem again, we find

$$\text{div}\,(\varepsilon_0 E + P) = \rho. \tag{1.13}$$

The electric displacement vector $D$ is used to denote the quantity $\varepsilon_0 E + P$ and (1.13) is the first of Maxwell's equations (1.1).

The internal fields in an insulator vary enormously on an atomic scale; however, the *average* response $P$ is usually linearly proportional to the applied field so there is some relation of the form

$$D_i = \sum_j \varepsilon_{ij} E_j, \qquad i = 1, 2, 3, \tag{1.14}$$

where $i$ is used to designate one of the components of the vector. $\varepsilon_{ij}$ is the *dielectric tensor* and the magnitude of its components depends on the nature of the insulator. For an isotropic material (1.14) simplifies to

$$D = \varepsilon E. \tag{1.15}$$

The *dielectric constant* $\varepsilon$ can be approximately calculated from an atomic model of the insulator; for our purposes it will be taken as a parameter of the theory to be determined separately, either by direct measurement or from a first principles calculation.

After switching on an electric field, there is a time lag before the polarization attains its full value because of the inertia of the electrons and nuclei. As a consequence, if a periodic field of frequency $\omega$ is applied, the dielectric constant will not in fact be constant, but vary as a function of $\omega$. In the extreme limit $\omega \to \infty$ the electrons get no opportunity to respond to the field which rapidly changes in direction, the polarization is zero, so in this limit in all insulators $\varepsilon \to \varepsilon_0$.

In conductors the electrons are free to move over macroscopic distances. In metals, which are good conductors, some of the outer shell atomic electrons are loosely bound and freely move within the metal. The movement of charge constitutes a *current* $J$ ($J = \sum_i q_i v_i$ for charges $q_i$ moving with velocity $v_i$). In gases and electrolytes, charged ions, which are atoms having an electron excess or deficiency, also make some contribution to the current. When the motion of the electrons is induced by an applied electric field there are various mechanisms which retard their motion. In most isotropic conductors, there is a simple linear relation between the applied field $E$ and the resultant current density $J$, namely

$$J = \sigma E, \tag{1.16}$$

known as Ohm's Law. The parameter $\sigma$ is the *electrical conductivity* of the conducting material. Equation (1.16) implies that the average velocity of the charges is proportional to the applied electric field. If the charges were not subjected to a retardation the acceleration and not the average velocity would be proportional to the applied field. The effectiveness of the scattering mechanisms that cause the retardation depends upon the temperature $T$ of the conductor and the frequency $\omega$ of the applied field, therefore $\sigma$ is a function of $T$ and $\omega$.

There are enormous differences in magnitude between the $\sigma$'s for good and bad conductors. Efficient conductors like the metals copper and silver have conductivities of the order of $10^7$ siemens/metre, relatively poor conductors like some liquids have $\sigma$ values of the order of $10^{-3}$ siemens/m, while very poor conductors like paraffin oils have conductivities of the order of $10^{-11}$ siemens/m.

From the principle of conservation of charge, a continuity equation relating $J$ and $\rho$ can be deduced. For a simple closed surface $S$ bounding a volume $V$, the rate of loss of charge from $V$ must equal that rate at which it flows through the surface so that

$$\int_S \mathbf{J} \cdot d\mathbf{S} = - \int_V \frac{\partial \rho}{\partial t} \, d\tau. \qquad (1.17)$$

By the use of the divergence theorem, and the fact that the equality is true for an arbitrary volume $V$, we obtain the differential form

$$\operatorname{div} \mathbf{J} + \frac{\partial \rho}{\partial t} = 0. \qquad (1.18)$$

The elimination of $\mathbf{J}$ between equations (1.18) and (1.16), and then $\mathbf{E}$ between (1.1) and (1·15), results in an equation which must be satisfied by the charge distribution function in a homogeneous conductor

$$\frac{\partial \rho}{\partial t} = - \frac{\sigma \rho}{\varepsilon}. \qquad (1.19)$$

It follows directly from this equation that there can be no permanent distribution of charge within a homogeneous material which has a finite conductivity. If $\rho_0$ is the charge distribution at $t = 0$, the appropriate solution of (1.19) for $\rho$ is

$$\rho(t) = \rho_0 e^{-\frac{\sigma t}{\varepsilon}}. \qquad (1.20)$$

The charge distribution decays exponentially and the charge eventually resides on the surface of the conductor. The time scale $\tau$ for the decay to occur is of the order of $\varepsilon/\sigma$. For metals $\tau$ can be extremely small, of the order of $10^{-19}$ seconds, for water $10^{-6}$ seconds and for very poor conductors minutes or even days. Whether a material is to be regarded primarily as a conductor or an insulator depends on the time scale of the phenomena under consideration. When fields of frequency $\omega$ are applied $\tau$ must be compared with the period of the field $2\pi/\omega$. A poor conductor will behave more like an insulator at very high frequencies, such that $\dfrac{\omega \varepsilon}{2\pi\sigma} \gg 1$, and like a conductor at low frequencies $\dfrac{\omega \varepsilon}{2\pi\sigma} \ll 1$.

## 1.4 The magnetic induction and magnetic field

When charges move with respect to one another, a further force comes into play. The total force $\mathbf{F}$ on a charge $q$ moving with a velocity $\mathbf{u}$ can be written in the form

$$\mathbf{F} = q(\mathbf{E} + \mathbf{u} \wedge \mathbf{B}), \qquad (1.21)$$

known as the Lorentz (1853–1928) force. The vector $B$ is the *magnetic induction* field and originates in the motion of other charges. The magnetic induction at a point $r$ due to a steady current density $J(r)$ flowing in a circuit which is completely enclosed in a volume $V$ is

$$B(r) = \frac{\mu_0}{4\pi} \int_V \frac{J(r') \wedge (r - r')d\tau'}{|r - r'|^3}. \tag{1.22}$$

This expression is based on the experimental law of Biot (1774–1862) and Savart (1791–1841) formulated in the early nineteenth century. It is usually given in the form of the field due to a small current element, but as this element must be a steady current in a complete circuit, we give the field in (1.22) due to the total current distribution. The proportionality constant $\mu_0$ is known as the permeability of free space and has the value $4\pi \times 10^{-7}$ henries/metre in S.I. units.

The divergence of (1.22) can be taken by differentiating under the integral sign to give

$$\text{div } B = 0, \tag{1.23}$$

which is the second of Maxwell's equations.

There is a generalization based on (1.22) generally known as Ampère's law. It was Ampère (1775–1836) and Oersted (1777–1851), Biot and Savart who made the early experimental investigations on the magnetic effects of currents. This law states that the total flux $\int J \cdot dS$ of a steady current through any surface $S$ bounded by a contour $C$ is equal to the line integral of $B$ round $C$ divided by $\mu_0$, i.e.

$$\int_S J \cdot dS = \frac{1}{\mu_0} \oint_C B \cdot dr. \tag{1.24}$$

By using Stokes's theorem, the line integral can be expressed as a surface integral of curl $B$ over $S$. The equality is true for any surface $S$ so (1.24) can be re-expressed in a differential form

$$\text{curl } B = \mu_0 J. \tag{1.25}$$

This result can be derived by direct differentiation of (1.22) using the the fact that div $J = 0$ $\left( \text{because } \dfrac{\partial \rho}{\partial t} = 0 \right)$ for a steady current.

The equation has to be modified within a material due to the polarization induced by the applied magnetic induction. In most substances, the induced fields are very small and, if they are such as to oppose the applied field, the material is classified as being *diamagnetic.* In so called *paramagnetics*, the effect is positive and larger as the atoms of these materials possess intrinsic magnetic moments which are partially aligned by the applied field. Below a certain critical temperature, paramagnetic substances can spontaneously order their atomic magnetic moments and produce large magnetic fields. In most substances the temperature at which this *ferromagnetic* phase sets in is extremely small, but a few substances, permanent magnetic materials such as iron and nickel, exist in this phase at room temperature.

The polarization is measured in terms of the induced magnetic dipole moment per unit volume $M$. This, it can be shown, is equivalent to a current distribution $J = \text{curl } M$. If this is taken into account, Ampère's law can be modified to include the magnetic polarization

$$\text{curl} \left( \frac{B}{\mu_0} - M \right) = J. \tag{1.26}$$

The magnetic field $H$ is defined as $\dfrac{B}{\mu_0} - M$ and (1.26) is written as

$$\text{curl } H = J. \tag{1.27}$$

In most materials, whether diamagnetic or paramagnetic, the polarization is proportional to the applied field. In isotropic substances,

$$B = \mu H, \tag{1.28}$$

where $\mu$ is known as the *permeability* of the material.

It is clear that (1.27) is valid only for steady currents because, by taking the divergence of both sides of the equation, we find

$$\text{div } J = 0, \tag{1.29}$$

which implies that $\dfrac{\partial \rho}{\partial t} = 0$.

Maxwell generalized the equation to apply to time dependent fields in what seemed to be the simplest way consistent with the charge continuity equation (1.18). If a vector $C$, which is to be determined, is added to the right-hand side of (1.27), then by taking the divergence of both sides, we find

$$\text{div } J + \text{div } C = 0. \tag{1.30}$$

From (1.18) and (1.13)

$$\text{div } C = \text{div } \frac{\partial D}{\partial t}, \tag{1.31}$$

where we have interchanged the order of the spatial and time derivatives. If $C$ is chosen to be $\dfrac{\partial D}{\partial t}$ we obtain the equation,

$$\operatorname{curl} \boldsymbol{H} = \boldsymbol{J} + \frac{\partial \boldsymbol{D}}{\partial t}, \tag{1.32}$$

which is consistent with the condition for charge continuity and is the third of Maxwell's equations. There have been attempts to justify this equation on more physical grounds and to interpret $\dfrac{\partial D}{\partial t}$ as a displacement current. As $\boldsymbol{D} = \varepsilon_0 \boldsymbol{E} + \boldsymbol{P}$, part of the term $\dfrac{\partial D}{\partial t}$ can be interpreted as a polarization current $\dfrac{\partial P}{\partial t}$ due to the change of the induced charge distribution with time. There are difficulties in interpreting $\varepsilon_0 \dfrac{\partial E}{\partial t}$ too literally as a current because a current generates a magnetic field, whereas the real source of both the electric and magnetic fields is in the charges of the system.

The last of Maxwell's equations (1.4) expresses Faraday's law of electromagnetic induction. Faraday argued that if a current could produce a magnetic field, then possibly a magnetic field could produce a current. This was confirmed, though not quite in the way anticipated, by his discovery that a current could be induced by a *change* in magnetic induction threading a circuit. Expressed quantitatively, the law states that the total electro-motive force $\oint \boldsymbol{E} \cdot d\boldsymbol{r}$ round a circuit $C$, which leads to an induced current in the circuit, is equal to the rate of decrease in magnetic flux through any surface bounded by $C$ so that

$$\oint_C \boldsymbol{E} \cdot d\boldsymbol{r} = - \frac{d}{dt} \int \boldsymbol{B} \cdot d\boldsymbol{S}. \tag{1.33}$$

By using Stokes's theorem (1.33) can be re-expressed in the differential form, if we make the assumption that this relation is valid for any contour $C$ and surface $S$, which is

$$\operatorname{curl} \boldsymbol{E} = - \frac{\partial \boldsymbol{B}}{\partial t}. \tag{1.34}$$

This completes the list of Maxwell's equations. They must be supplemented by the relations between $\boldsymbol{D}$ and $\boldsymbol{E}$ (1.15) and between

*B* and *H* (1.28). For a complete set of dynamical principles the Lorentz equation (1.21) for the force on a charge due to the fields *E* and *B* must be added to enable the equation of motion of the charge to be calculated. In a conductor, Ohm's Law (1.16) can be used instead, as this plays the role of an average equation of motion of charge within a conductor.

The evidence for Maxwell's equations was based on experiments examining the behaviour of systems of static and quasi-static charges and currents. There is little direct evidence to justify the form of the individual equations for rapidly fluctuating charge and current densities and high frequency fields. Their justification in this case is based on the experimental confirmation of predictions based on all four equations. The displacement current plays an important role, and it is due to this term that a wave equation can be deduced. Considerable experimental information on the behaviour of electro-magnetic waves has been accumulated, which agrees in detail with explanations based on these equations; this gives a convincing amount of evidence that the equations are of the correct form to describe high frequency fields.

## 1.5 Conservation theorems

Charge continuity, and hence charge conservation, is expressed in equation (1.18) which is built into Maxwell's equations via the displacement current. If Maxwell's equations are taken as one's starting-point, (1.18) can be deduced from (1.3) and (1.1), and can be interpreted as implying charge conservation.

An identity can be deduced which can be interpreted as expressing the conservation of energy. Most treatises on electro- and magneto-statics prove that the total increase in energy of a medium in a volume $V$ due to small increases in the electric and magnetic fields of $\delta D$ and $\delta B$ is

$$\int_V (E \cdot \delta D + H \cdot \delta B)\,dt, \tag{1.35}$$

if it is assumed that the parameters characterising the medium are unaffected by the increases. If it is assumed that (1.35) holds for time dependent fields, the rate of change of electromagnetic energy stored in $V$ is

$$\int_V \left( E \cdot \frac{\partial D}{\partial t} + H \cdot \frac{\partial B}{\partial t} \right) d\tau. \tag{1.36}$$

From (1.3) and (1.4) we deduce

$$E \cdot \frac{\partial D}{\partial t} + H \cdot \frac{\partial B}{\partial t} = -(H \cdot \mathrm{curl}\ E - E \cdot \mathrm{curl}\ H) - E \cdot J. \quad (1.37)$$

The first two terms on the right-hand side of this equation in brackets are an expansion of div $(E \wedge H)$, so by integrating (1.37) throughout a volume $V$ we obtain the identity

$$\int_V \left( E \cdot \frac{\partial D}{\partial t} + H \cdot \frac{\partial B}{\partial t} \right) d\tau = - \int_V \mathrm{div}\ (E \wedge H) d\tau - \int_V E \cdot J d\tau. \quad (1.38)$$

If the volume $V$ is bounded by a surface $S$, then the first integral on the right-hand side can be re-expressed as a surface integral of $E \wedge H$ over $S$ by the use of the divergence theorem, whence

$$\int_V \left( E \cdot \frac{\partial D}{\partial t} + H \cdot \frac{\partial B}{\partial t} \right) d\tau = \int_S (E \wedge H) \cdot dS + \int_V E \cdot J d\tau. \quad (1.39)$$

The rate at which work is done by fields $E$ and $B$ on a charge $q$ which moves with a velocity $u$ is $F \cdot u$, where $F$ is the Lorentz force (1.21). The magnetic field does no work as it always acts in a direction perpendicular to the velocity of the charge, so the rate of working is $qE \cdot u$. For an assembly of charges this is $E \cdot J$ per unit volume. Using Ohm's Law, $E \cdot J$ is equal to $J^2/\sigma$, which is the energy dissipated in the form of heat in the conductor (Joule heating). If the left-hand side of (1.39) is interpreted as the rate of decrease of electromagnetic energy stored in $V$, the last term as the heat generated in $V$, the term $\int_S (E \wedge H) \cdot dS,$ can be interpreted as the rate of energy flow through $S$. The product $E \wedge H$ is known as Poynting's vector $S$. It is tempting to interpret it as a measure of the intensity of the electromagnetic energy flow at a point. However, any vector whose divergence is zero can be added to it without affecting (1.39). Static fields $E$ and $H$ at right-angles give a finite Poynting's vector but are not thought to involve any energy flow. Only the integrated form of Poynting's vector over a closed surface can be interpreted as an energy flow, however, the interpretation of $S$ as the electromagnetic intensity does not often lead to difficulties.

Formula (1.39) is the usual form for expressing energy conservation and is the most useful, though there are other ways in which the equation can be separated and interpreted.

## 1.6 Boundary conditions

The usual requirement of a mathematical problem which is to be used to describe some physical system is that it must have a unique solution. For Maxwell's equations this means specifying enough initial and boundary conditions on the fields to ensure uniqueness. To this end, we shall now deduce some conditions on the fields at the boundary between two media. The interface between two media is usually a very narrow transition region in which the local constants $\varepsilon$, $\mu$, $\sigma$ change very rapidly from the bulk values in one medium $\varepsilon_1$, $\mu_1$, $\sigma_1$, to the bulk values in the second medium $\varepsilon_2$, $\mu_2$, $\sigma_2$. This region is generally so small on a macroscopic scale that the change in properties can be taken to be discontinuous. If Maxwell's equations hold over the boundary region, they can be used to deduce expressions relating the fields on either side of the boundary.

Consider a cylinder with a small cross-section of area $A$ and length $l$ with its axis normal to the surface, one end face in medium one and the other in medium two (see Fig. 1 (a)). By using the divergence theorem, (1.2) can be expressed in the integral form

$$\oint_S \boldsymbol{B} \cdot d\boldsymbol{S} = 0. \tag{1.40}$$

This means that the total magnetic flux through any closed surface must be zero. If this is applied to the cylinder, and the variation of $\boldsymbol{B}$ over the end faces is neglected because $A$ can be chosen to be as small as one likes, then we have

$$(\boldsymbol{B}_1 \cdot \boldsymbol{n} - \boldsymbol{B}_2 \cdot \boldsymbol{n})A + \text{contribution from curved surface} = 0, \tag{1.41}$$

where $\boldsymbol{n}$ is a unit vector normal to the interface, and the subscripts denote the medium to which fields correspond. In the limit as $l \rightarrow 0$ the contribution from the curved surface is zero, so in this limit we find

$$\boldsymbol{B}_1 \cdot \boldsymbol{n} = \boldsymbol{B}_2 \cdot \boldsymbol{n}, \tag{1.42}$$

which states that the normal component of $\boldsymbol{B}$ is continuous over the interface.

Similarly (1.1) can be written in the integral form,

$$\int_S \boldsymbol{D} \cdot d\boldsymbol{S} = Q, \tag{1.43}$$

where $Q$ is the total charge within the surface $S$. If this is applied to the cylinder, in the limit $l \to 0$

$$(\mathbf{D}_1 \cdot \mathbf{n} - \mathbf{D}_2 \cdot \mathbf{n})A = \lim_{l \to 0}\left(\frac{Q}{l}\right). \tag{1.44}$$

The right-hand side can be finite because there can be an infinitesimal layer of charge (on a macroscopic scale) at the interface. If $\rho_s$ is used to denote the surface charge density per unit area,

$$(\mathbf{D}_1 - \mathbf{D}_2) \cdot \mathbf{n} = \rho_s. \tag{1.45}$$

There is a discontinuity in the normal component of the electric displacement equal to the surface charge density at the interface.

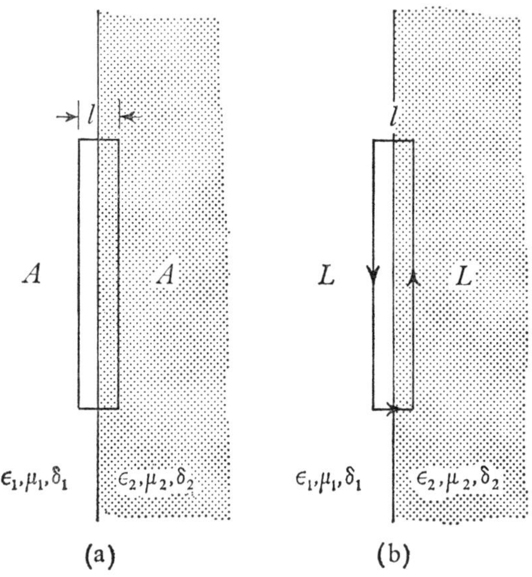


Fig. 1 (a) represents a cylinder of cross-sectional area $A$ and of length $l$ through the interface between medium 1 and medium 2; (b) represents a rectangular contour of dimensions $L \times l$ through the same interface

To use the integral form of equations (1.3) and (1.4), we choose a small rectangular contour which passes through the interface (see Fig. 1 (b)). Let the two sides of length $l$ be along the normal. The integral form of (1.4) is

$$\int_C \mathbf{E} \cdot d\mathbf{r} = -\int_S \frac{\partial \mathbf{B}}{\partial t} \cdot d\mathbf{S}, \tag{1.46}$$

where $S$ is a surface bounded by the contour $C$. If (1.46) is applied to the rectangle, and the variation of $E$ along the side of length $l$ is ignored, then in the limit $l \to 0$ we can show that

$$E_1 \wedge n = E_2 \wedge n \tag{1.47}$$

if $\dfrac{\partial B}{\partial t}$ is bounded. This means that the tangential component of the electric field is continuous over the interface.

Similarly, using the integral version of (1.3),

$$\int_C H \cdot dr = \int_S J \cdot dS + \int_S \frac{\partial D}{\partial t} \cdot dS, \tag{1.48}$$

we can deduce that the tangential component of the magnetic field is also continuous over the interface, that is

$$H_1 \wedge n = H_2 \wedge n. \tag{1.49}$$

This assumes that the $\dfrac{\partial D}{\partial t}$ and $J$ are bounded. The latter is true if $\sigma_1$ and $\sigma_2$ are finite, but it is sometimes useful to approximate a good conductor by a perfect conductor ($\sigma \to \infty$). If medium two is a perfect conductor, the current density need not be bounded and a surface current $J_s$ can exist at the interface. In this case the tangential components of the magnetic field are discontinuous over the interface and are related by the equation

$$n \wedge (H_1 - H_2) = J_s. \tag{1.50}$$

## PROBLEMS 1

1. If $E(r, t) = \mathrm{Re}\, E_0(r)e^{i\omega t}$, $H(r, t) = \mathrm{Re}\, H_0(r)e^{i\omega t}$ show that the time average of the Poynting's vector is equal to the real part of $\frac{1}{2}E_0 \wedge H_0^*$, where $H_0^*$ is the complex conjugate of $H_0$.

**CHAPTER 2**

# Electromagnetic Waves

## 2.1 The vector wave equation

For a preliminary investigation of Maxwell's equations we restrict the discussion to the equations for free space ($\varepsilon = \varepsilon_0$, $\mu = \mu_0$) in the absence of charge or current sources. In this case the equations simplify to

$$\text{div } \mathbf{D} = 0 \quad (2.1), \qquad \text{div } \mathbf{B} = 0 \quad (2.2),$$

$$\text{curl } \mathbf{H} = \frac{\partial \mathbf{D}}{\partial t} \quad (2.3), \qquad \text{curl } \mathbf{E} = -\frac{\partial \mathbf{B}}{\partial t}. \quad (2.4)$$

As $\mathbf{D} = \varepsilon_0 \mathbf{E}$, $\mathbf{B} = \mu_0 \mathbf{H}$, the equations can be written in terms of the two fields $\mathbf{E}$ and $\mathbf{H}$ only. From (2.3) and (2.4) we can deduce equations involving only one of the fields

$$\text{curl curl } \mathbf{E} = -\varepsilon_0 \mu_0 \frac{\partial^2 \mathbf{E}}{\partial t^2}, \tag{2.5}$$

$$\text{curl curl } \mathbf{H} = -\varepsilon_0 \mu_0 \frac{\partial^2 \mathbf{H}}{\partial t^2}, \tag{2.6}$$

where it has been assumed that the partial derivatives of the fields are continuous functions of the time and space variables so that the order of the differentiation can be reversed.

These equations can be re-written by using the vector operator $\nabla^2$ defined by

$$\nabla^2 \mathbf{A} = \text{grad div } \mathbf{A} - \text{curl curl } \mathbf{A}, \tag{2.7}$$

where $\mathbf{A}$ is an arbitrary vector. In cartesian coordinates

$$\nabla^2 \mathbf{A} = i\nabla^2 A_1 + j\nabla^2 A_2 + k\nabla^2 A_3,$$

where $i, j, k$ are the cartesian unit vectors, $A_1, A_2, A_3$ are the cartesian components of $\mathbf{A}$, and $\nabla^2$ is the scalar operator nabla. For other coordinate systems $\nabla^2 \mathbf{A}$ takes a more complicated form because each component of $\nabla^2 \mathbf{A}$ usually involves the derivatives of all three components of $\mathbf{A}$. Using (2.1), (2.2) and (2.7) the equations (2.5), (2.6)

16

can be re-expressed in the form

$$\nabla^2 E = \mu_0 \varepsilon_0 \frac{\partial^2 E}{\partial t^2}, \tag{2.8}$$

$$\nabla^2 H = \mu_0 \varepsilon_0 \frac{\partial^2 H}{\partial t^2}, \tag{2.9}$$

which are the homogeneous vector wave equations. The reader is probably familiar with the scalar wave equation

$$\nabla^2 \phi = \frac{1}{v^2} \frac{\partial^2 \phi}{\partial t^2}, \tag{2.10}$$

which describes waves characterised by a scalar function $\phi$ and propagated with a velocity $v$. The general solution of the one-dimensional equation of the form (2.10) is $\phi(z, t) = f(z-vt) + g(z-vt)$, where $f(z)$ and $g(z)$ are arbitrary wave profiles that are propagated in the positive $z$ and negative $z$ directions respectively.

Clearly (2.8) and (2.9) describe the propagation of some form of electromagnetic wave with a velocity $c$ equal to $1/\sqrt{\varepsilon_0 \mu_0}$. This is equal to the velocity of light $2 \cdot 997 \times 10^8$ metres/second. However, as was mentioned at the beginning of Chapter 1, visible light covers only a very small part of the electromagnetic spectrum from $3 \cdot 9 \times 10^{-7}$ to $7 \cdot 7 \times 10^{-7}$ metres. The human species is not equipped with such a sensitive detector as the eye for other parts of the electromagnetic spectrum. These are detected indirectly, radio waves via a radio receiver which re-transmits the energy as sound waves, X-rays are detected by their effect on photographic plates, $\gamma$-rays by the electronic transitions they induce which result in the emission of waves in the visible spectrum.

## 2.2 Plane waves

It will be convenient to consider waves of a single frequency $\omega$; we take the electric and magnetic fields $E(r, t)$, $H(r, t)$ to be of the form the real part of $E_c(r)e^{i\omega t}$ and the real part of $H_c(r)e^{i\omega t}$, respectively. These fields correspond to a wave which is simple harmonic in time with a period $2\pi/\omega$. Fields with a more general time dependence will be considered later. From (2.8) and (2.9) we can deduce that these monochromatic waves must satisfy the equation

$$(\nabla^2 + \varepsilon_0 \mu_0 \omega^2)\begin{cases} E_c \\ H_c \end{cases} = 0. \tag{2.11}$$

For the simple case in which $E_c$ and $H_c$ depend upon one spatial coordinate only, say $z$, (2.11) becomes

$$\left(\frac{d^2}{dz^2} + \varepsilon_0\mu_0\omega^2\right)\begin{Bmatrix} E_c \\ H_c \end{Bmatrix} = 0. \tag{2.12}$$

The solutions of this equation in exponential form are

$$E_c(z) = E_0 e^{\pm i\omega\sqrt{\varepsilon_0\mu_0}\,z}, \quad H_c(z) = H_0 e^{i\omega\sqrt{\varepsilon_0\mu_0}\,z}. \tag{2.13}$$

These are plane wave solutions of the wave equation because the field amplitudes $E_c(r)$ and $H_c(r)$, are constant over planes perpendicular to the $z$-direction.

The fact that the solutions (2.13) satisfy the wave equation is not a sufficient condition for them to be solutions of Maxwell's equations. Conditions on $E_0$ and $H_0$ are obtained by requiring (2.13) to be solutions of equations (2.1) to (2.4). Substituting (2.13) into (2.1) gives

$$E_{0z} = 0, \quad H_{0z} = 0, \tag{2.14}$$

provided $\omega \neq 0$.

Substituting (2.13) into (2.3) or (2.4) gives the condition

$$H_0 = \mp \frac{k \wedge E_0}{\mu_0\omega}, \tag{2.15}$$

where $k$ is a vector in the $z$-direction of magnitude $\omega\sqrt{\varepsilon_0\mu_0}$.

These plane wave solutions can be more generally expressed in the form,

$$E(r, t) = E_0 e^{i(\omega t - k \cdot r)}, \tag{2.16}$$

$$H(r, t) = \frac{k \wedge E_0}{\mu_0\omega}\, e^{i(\omega t - k \cdot r)}. \tag{2.17}$$

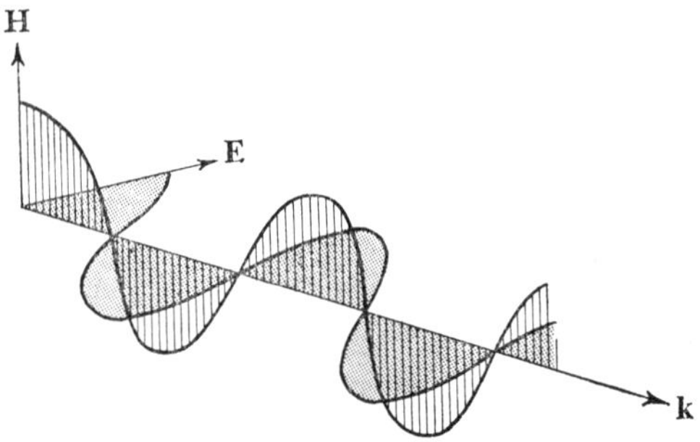


Fig. 2 A plane harmonic electromagnetic wave in free space

The energy associated with these waves is propagated in the direction of the Poynting vector $E \wedge H$ which, in this case, corresponds to the direction of the vector $k$ because the three vectors $E$, $H$, and $k$, form a mutually perpendicular triad. The electric and magnetic vectors of the wave are also in phase. These are features which are not true for all electromagnetic waves, as we shall see later.

## 2.3 Polarization

The plane wave solution (2.16), (2.17), where $E_0$ is a constant vector fixed in direction, is a *plane polarized* wave. There is some confusion in the literature in the use of the term *plane of polarization*; we shall use it to refer to the plane containing the vectors $E$ and $k$, though in optics it is often used to refer to the plane containing $H$ and $k$.

As a consequence of the linearity of Maxwell's equation, a linear combination of two plane waves with their planes of polarization at right angles

$$E = E_{0x}e^{i(\omega t - k \cdot r)} + E_{0y}e^{i(\omega t - k \cdot r)} \tag{2.18}$$

is also a solution of the equations. The resultant wave will not, strictly speaking, be monochromatic because the electric field amplitude $E_0$ will be time dependent. If there is a constant phase difference between the two component waves, the resultant amplitude $E_0$ will rotate with an angular velocity $\omega$ and its end point will trace out an ellipse. Such a wave is described as being *elliptically polarized*. A special case occurs when the amplitudes of the component waves are equal ($| E_{0x} | = | E_{0y} |$) and differ in phase by $\pi/2$, $E$ then traces out a circle and the wave is *circularly polarized*.

## 2.4 Dispersive media

We noted earlier that the dielectric constant and permeability depend upon the frequency of the applied fields due to the time lag of the induced polarization (which is not to be confused with the polarization of a wave referred to in 2.3). This effect causes waves of different frequencies to be propagated at different speeds; a phenomenon known as *dispersion*. If this time lag is ignored and the response is taken to be instantaneous then $D(t) = \varepsilon E(t)$, $B(t) = \mu H(t)$ and the wave equations in the form (2.8)(2.9) can be derived with $\varepsilon\mu$ replacing $\varepsilon_0\mu_0$. The velocity of these waves is $1/\sqrt{\varepsilon\mu}$, which must always be less than the velocity of light in free space.

A wave that is propagated through a dielectric comes from two sources. One is the original plane wave of the form (2.16) (2.17) which falls on the dielectric; the other is from secondary waves that are set up due to the polarization that this wave induces. The dielectric constant for an applied field of the form (2.16) is not only a function of the frequency $\omega$ of the field, but is also complex of the form

$$\varepsilon(\omega) = \varepsilon'(\omega) + i\varepsilon''(\omega). \tag{2.19}$$

The imaginary part is due to absorption of energy by the atoms of the dielectric from the applied field. The absorption is particularly marked, and displays a resonance form, when the frequency of the applied field coincides with the frequency of a mode of excitation of the atoms of the dielectric. When the absorption is strong, the real part of the dielectric constant varies rapidly as a function of $\omega$ (see Fig. 3). This very marked $\omega$-dependence is usually known as *anomalous dispersion*.

Between the regions of strong absorption there are transparent regions in which $\varepsilon''(\omega)$ is small and $\varepsilon'(\omega)$ varies slowly as a function of $\omega$. Within these transparent bands it may be a reasonable approximation to neglect $\varepsilon''(\omega)$ and take $\varepsilon'(\omega)$ as a constant. The permeability $\mu$ is also a function of $\omega$ and complex but, because the magnetic polarization is usually small compared to the electric polarization, it seldom has to be taken into account.

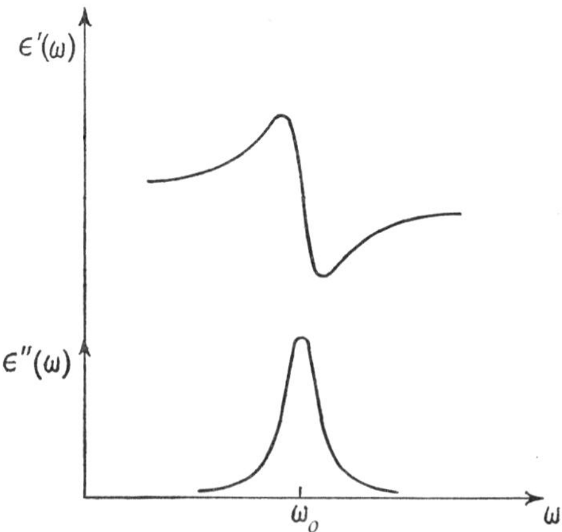


Fig. 3 A typical plot of $\epsilon'(\omega)$ and $\epsilon''(\omega)$ as a function of $\omega$ in the region of an absorption band about the excitation frequency $\omega_0$

When there is dispersion the wave equations in the form (2.8) (2.9) cannot be derived. However, for fields which have a single frequency component $E_c(r)e^{i\omega t}$, $H_c(r)e^{i\omega t}$, we can derive a wave equation in the form (2.11) by working with the Fourier transforms of the fields. For an electric vector $E(t)$ the Fourier transform $E_\omega$ is defined by

$$E_\omega = \frac{1}{\sqrt{2\pi}} \int_{-\infty}^{\infty} E(t)e^{-i\omega t}dt, \tag{2.20}$$

provided the integral is defined. The inverse transformation is

$$E(t) = \frac{1}{\sqrt{2\pi}} \int_{-\infty}^{\infty} E_\omega e^{i\omega t}d\omega. \tag{2.21}$$

This expresses the field $E(t)$ as a linear combination of fields over a continuous frequency spectrum $\omega$ with an amplitude distribution $E_\omega$.

If we take the Fourier transform of all Maxwell's equations (2.1) to (2.4) we obtain equations relating the Fourier transforms corresponding to the same frequency namely,

$$\operatorname{div} D_\omega = 0, \qquad (2.22), \qquad \operatorname{div} B_\omega = 0, \qquad (2.23)$$

$$\operatorname{curl} H_\omega = i\omega D_\omega, \qquad (2.24), \qquad \operatorname{curl} E_\omega = -i\omega B_\omega, \qquad (2.25)$$

where we have used the fact that the Fourier transform of $\dfrac{\partial D}{\partial t}$ is $i\omega D_\omega$.

When the lag of the polarization is taken into account, there is a more general relation between the fields $E$ and $D$,

$$D(t) = \int_{-\infty}^{0} \varepsilon(t-t')E(t')dt', \tag{2.26}$$

and a similar relation between $B$ and $H$. The Fourier transforms of these equations are

$$D_\omega = \varepsilon(\omega)E_\omega, \quad B_\omega = \mu(\omega)H_\omega, \tag{2.27}$$

where $\varepsilon(\omega)$, $\mu(\omega)$ are the Fourier transform of $\varepsilon(t)$ and $\mu(t)$.

The arguments used in section (2.2) can be used to derive an equation corresponding to (2.11), which is

$$(\nabla^2 + \varepsilon(\omega)\mu(\omega)\omega^2)\begin{cases} E_\omega \\ H_\omega \end{cases} = 0. \tag{2.28}$$

If $\varepsilon$ and $\mu$ are taken to be real, the plane wave solutions of this equation are

$$E_\omega(r) = E_0(\omega)e^{i(\omega t - k \cdot r)}, \tag{2.29}$$

$$H_\omega(r) = \frac{k \wedge E_0(\omega)}{\mu\omega} e^{i(\omega t - k \cdot r)}, \tag{2.30}$$

where $|k| = \omega\sqrt{\varepsilon(\omega)\mu(\omega)}$. The time dependence of the fields corresponding to these solutions can be calculated using the Fourier inversion (2.21). The solutions are arbitrary to the extent that the frequency dependence of $E_0(\omega)$ is arbitrary.

The surfaces of constant phase for the solutions for a single frequency component (2.29) (2.30) are planes perpendicular to the wave vector $k$ which travel with a velocity $1/\sqrt{\varepsilon(\omega)\mu(\omega)}$, known as the phase velocity. When a wave form composed of several frequency components is propagated in a dispersive medium, the wave is distorted, because the different phase velocities cause the distribution of phases to vary as the wave travels. When this occurs, one cannot define the velocity of a wave profile. In a non-absorbing medium it is possible to define the average velocity of the energy flow but this can be done in several ways. One way is to use the concept of a *group velocity* $v_g$. A small energy excess is created by combining two waves of slightly different frequencies $\omega_1 = \omega + \Delta\omega$, $\omega_2 = \omega - \Delta\omega$. If the electric vector of the waves is of the form $E_0 \sin (kz - \omega t)$, the electric vector of the total wave is

$$E = E_0\{\sin [k(\omega_1)z - \omega_1 t] + \sin [k(\omega_2)z - \omega_2 t]\}. \tag{2.31}$$

If $\Delta\omega$ is very small (2.31) can be written approximately as

$$E = E_0 \cos \left[ \Delta\omega \left( \frac{\partial k}{\partial \omega} z - t \right) \right] \sin [k(\omega)z - \omega t]. \tag{2.32}$$

This describes a sine wave of frequency $\omega$ with an amplitude modulated by the cosine term. The group velocity is the velocity of the modulated amplitude or energy excess. From (2.32) we see

$$v_g = \frac{\partial \omega}{\partial k}. \tag{2.33}$$

When there is no dispersion $\varepsilon$ and $\mu$ are independent of $\omega$ and the group velocity is the same as the phase velocity. In three dimensions the group velocity is

$$v_{gi} = \frac{\partial \omega}{\partial k_i}, \qquad i = 1, 2, 3, \tag{2.34}$$

where $(k_1, k_2, k_3)$ is the propagation vector $\mathbf{k}$ and $(v_{g_1}, v_{g_2}, v_{g_3})$ is the group velocity $\mathbf{v}_g$ in cartesian coordinates.

Another way of defining the velocity of energy transport is by the average rate of energy propagation divided by the average energy, a quantity we shall denote by $v_E$. For a plane wave with an electric vector $\mathbf{E} = \mathbf{E}_0 \sin (kz - \omega t)$ the average rate of energy flow can be calculated from the average of the Poynting vector over a time interval $2\pi/\omega$. This is

$$\bar{S} = \sqrt{\frac{\varepsilon}{\mu}} \frac{|E_0|^2}{2}. \tag{2.35}$$

The computation of the average energy density

$$U = \int \mathbf{E} \cdot d\mathbf{D} + \mathbf{H} \cdot d\mathbf{B}, \tag{2.36}$$

is rather more difficult. When there is dispersion, the energy is not just equal to the energy differences of the wave between the field values $E = 0, E = E_0$ because the polarization lags behind the applied field and the electrons of the dielectric have kinetic energy due to the wave when the electric field value has fallen to zero. This difficulty can be avoided by building up the wave infinitesimally slowly from two waves of slightly different frequencies $\omega_1 = \omega + \Delta\omega$, $\omega_2 = \omega - \Delta\omega$, which are out of phase by $\pi$ at $t = 0$. When $\omega$ is small the resultant wave builds up slowly from zero to a wave of frequency $\omega$ and amplitude $E_0$ in a time $\pi/2\Delta\omega$. The electric vector of the wave is

$$E(z, t) = \frac{E_0}{2} \{\sin [k(\omega_1)z - \omega_1 t] - \sin [k(\omega_2)z - \omega_2 t]\} \tag{2.37}$$

and its displacement vector is

$$D(z, t) = \frac{E_0}{2} \{\varepsilon(\omega_1) \sin [k(\omega_1)z - \omega_1 t] - \varepsilon(\omega_2) \sin [k(\omega_2)t - \omega_2 t]\}. \tag{2.38}$$

Once the corresponding magnetic fields have been calculated the total energy required to create the wave of frequency $\omega$ can be found by evaluating

$$\bar{U} = \underset{\Delta\omega \to 0}{\text{limit}} \int_0^{\frac{\pi}{2\Delta\omega}} \left( \boldsymbol{E} \cdot \frac{\partial \boldsymbol{D}}{\partial t} + \boldsymbol{H} \cdot \frac{\partial \boldsymbol{B}}{\partial t} \right) dt. \tag{2.39}$$

If the $\omega$ dependence of $\mu$ is neglected, it is not too difficult to show that the integral (2.39) has the value

$$\bar{U} = |\, \boldsymbol{E}_0 \,|^2 \left( \varepsilon + \omega \frac{\partial \varepsilon}{\partial \omega} \right). \tag{2.40}$$

From (2.35) and (2.40) we can find the average velocity of energy transport. It is given by

$$\frac{1}{v_E} = \tfrac{1}{2} \sqrt{\mu/\varepsilon} \left( \varepsilon + \frac{\partial \varepsilon}{\partial \omega} \right), \tag{2.41}$$

This expression for $v_E$ is identical to the group velocity $v_g$ given by (2.33). It can be shown that these two velocities are equal when the $\omega$-dependence of $\mu$ is also taken into account.

There is yet another way of defining the velocity of energy transport which is known as the *signal velocity*. It is calculated from the time taken for a semi-infinite sinusoidal wave of frequency $\omega$ to reach a sensitive detector. This problem has been considered in detail by Sommerfeld and Brillouin (see [1]†), with some interesting results. If the detector can respond to arbitrarily small fields, the wave front will be observed to travel with the velocity of light in *free space* $1/\sqrt{\varepsilon_0\mu_0}$. The amplitude builds up as secondary waves are established due to the forced oscillations of the electrons in the dielectric. If the detector is made realistic and required to have a finite threshold for the field values that it can detect, the observed velocity will depend upon its sensitivity. If the sensitivity threshold is fixed at half the amplitude of the final waveform, the signal velocity deduced is equal to $v_E$ and $v_g$, provided there is no absorption.

To be consistent with the theory of relativity the velocity of the energy flow cannot exceed the velocity of light in free space $c$ (this does not apply to the phase velocity). In the regions of anomalous dispersion where $\varepsilon$ is negative, the group velocity can be greater than $c$. However, in these regions there must be strong absorption, the imaginary part of $\varepsilon(\omega)$ cannot be ignored, and the group velocity cannot be interpreted as the velocity of energy flow.

† Numbers in bold, e.g. [1], refer to the bibliography at the end of the book.

## 2.5 Absorption

There are several mechanisms by which energy can be absorbed
from an electromagnetic wave. In a dielectric, $\varepsilon(\omega)$ and $\mu(\omega)$ have
imaginary parts because energy is absorbed through some form of
excitation of the dielectric atoms. In a conductor energy is absorbed
in the scattering mechanisms which retard the electrons, causing the
temperature of the conductor to rise. In terms of the Fourier transforms of the electric field and current density, Ohm's Law is

$$J(\omega) = \sigma(\omega)E_\omega. \tag{2.42}$$

If the current density $J$ is retained in (1.3), (2.28) is generalized to

$$\left[\nabla^2 + \omega^2\varepsilon\mu\left(1 - \frac{i\sigma}{\omega\varepsilon}\right)\right]\begin{Bmatrix} E_\omega \\ H_\omega \end{Bmatrix} = 0. \tag{2.43}$$

The solutions (2.29) (2.30) are also solutions of (2.43) and Maxwell's
equations, provided $k$ is redefined as a complex vector $k' - ik''$, such
that

$$k'^2 - k''^2 - 2ik' \cdot k'' = \omega^2\varepsilon\mu\left(1 - \frac{i\sigma}{\omega\varepsilon}\right). \tag{2.44}$$

The relative direction of the two vectors $k'$ and $k''$ is undefined. If it
is given, the magnitude of the vectors can be calculated from (2.44)
by separating it into its real and imaginary parts. The imaginary
parts of $\varepsilon$ and $\mu$ contribute to $k''$ together with the conductivity $\sigma$.

When $k$ is complex, we see from (2.29) (2.30) that the wave is
attenuated by a factor $e^{-k'' \cdot r}$. Such a wave is in general not a plane
wave because the phase of the wave is constant over phase perpendicular to $k'$ while the amplitude is constant over planes perpendicular to
$k''$; only when these directions coincide does it become a plane wave.
There is a phase difference between the electric and magnetic vectors.
This difference is equal to $\tan^{-1}\left(\dfrac{|k''|}{|k'|}\right)$ when $k'$ and $k''$ are in the
same direction.

A good approximation can be made in very good conductors for
waves with frequencies less than optical frequencies such that $\dfrac{\varepsilon\omega}{\sigma} \ll 1$.
The first term on the right-hand side of (2.44) can be neglected in
comparison with the second. If $k'$ and $k''$ are in the same direction,
and $\mu$ and $\varepsilon$ are real, (2.44) can be solved for $|k'|$, $|k''|$;

$$|k'| \approx |k''| \quad \text{and} \quad |k''| \approx \sqrt{\frac{\mu\omega\sigma}{2}}. \tag{2.45}$$

In this approximation the phase difference between the electric and magnetic fields is $\pi/4$. The amplitude of the fields falls off by a factor $1/e$ after a distance $\sqrt{2/\mu\sigma\omega}$. In a typical situation, such as micro-frequency wave $\omega \sim 10^{10}$ cps, in a good conductor like silver $\sigma \simeq 3 \times 10^{10}$ sie/m, $\sqrt{2/\mu\sigma\omega}$ is approximately $10^{-4}$ cm. This means that the current carried in silver at these frequencies is confined to a very thin surface layer. In the limit of a perfect conductor $\sigma \to \infty$ the current is purely a surface one.

## 2.6 Anisotropic Media

In crystals the relation between $D$ and $E$ cannot be characterized by a single constant $\varepsilon$; the more general expression involving the dielectric tensor (1.14) is required. There are only six independent elements in the dielectric tensor because three are determined by the condition of symmetry $\varepsilon_{ij} = \varepsilon_{ji}$. Because of the symmetry, axes can always be chosen so that the tensor is diagonal

$$D_i = \varepsilon_i E_i. \quad i = 1, 2, 3. \tag{2.46}$$

In a uniaxial crystal two of the diagonal components are equal, $\varepsilon_1 = \varepsilon_2 = \varepsilon_\perp$, $\varepsilon_3 = \varepsilon_\parallel$, where the third axis is taken to be along the axis of symmetry of the crystal. In biaxial crystals, the diagonal elements are distinct and the set of axes for which the dielectric tensor is diagonal bears no simple relation to the axes of the crystal and can depend upon the frequency of the applied field if this alters the relative values of the elements of the tensor.

Plane wave solutions of the form

$$E = E_0 e^{i(\omega t - k \cdot r)}, \quad H = H_0 e^{i(\omega t - k \cdot r)}, \tag{2.47}$$

satisfy Maxwell's equation if

$$k \cdot H_0 = 0, \qquad k \cdot D_0 = 0, \tag{2.48}$$

$$\omega D_0 = -k \wedge H_0, \quad \omega\mu H_0 = k \wedge E_0, \quad | \tag{2.49}$$

$k$, $D_0$ and $H_0$ form a mutually perpendicular triad, and $E_0$, $D_0$ and $k$ are coplanar. Eliminating $H_0$ between (2.48) and (2.49) gives a vector equation relating $D$ and $E$,

$$D_0 = \frac{k \wedge (E_0 \wedge k)}{\mu\omega^2}. \tag{2.50}$$

The components of $D$ can be substituted from (2.46) into (2.50) in terms of the components of $E$ to give three homogeneous equations

$$\mu\omega^2 \varepsilon_i E_{oi} = k^2 E_{oi} - (k \cdot E_o)k_i, \quad i = 1, 2, 3. \tag{2.51}$$

These equations have a non-trivial solution for the components of $E$ only if the determinant of the coefficients is zero. This condition can be put in the form,

$$\frac{k_1^2}{k^2 - \mu\varepsilon_1\omega^2} + \frac{k_2^2}{k^2 - \mu\varepsilon_2\omega^2} + \frac{k_3^2}{k^2 - \mu\varepsilon_3\omega^2} = 1, \qquad (2.52)$$

which is Fresnel's equation, in which $(k_1, k_2, k_3)$ are the cartesian components of the vector $k$. If the direction of $k$ is given, then (2.52) is a quadratic equation for $k^2$ which will in general have two solutions corresponding to two waves travelling with different phase velocities.

In uniaxial crystals (2.52) factorizes into the two equations

$$k_1^2 + k_2^2 + k_3^2 = \omega^2\mu\varepsilon_{\parallel}, \qquad (2.53)$$

and

$$\frac{k_1^2 + k_2^2}{\mu\varepsilon_{\parallel}\omega^2} + \frac{k_3^2}{\mu\varepsilon_{\perp}\omega^2} = 1. \qquad (2.54)$$

the plane wave corresponding to the solution (2.53) has a phase velocity $1/\sqrt{\mu\varepsilon_{\parallel}}$, independent of the direction of $k$. Substituting $k^2$ from (2.53) into (2.51) for $i = 3$, we find $k \cdot E_0 = 0$, which means that $E_0$ is perpendicular to $k$, and from (2.50) that $D_0$ and $E_0$ are in the same direction. If this is to be compatible with (2.46), $D_0$ and $E_0$ must be perpendicular to the axis of the crystal. This wave, which is similar to the plane wave in an isotropic dielectric, is known as the *ordinary* wave. It is plane polarized with its electric vector perpendicular to the plane containing $k$ and the crystal axis, a plane that is known as the *principal section*.

In contrast, the solution of (2.54) for $k^2$ depends upon the direction of $k$. The wave it describes is known as the *extraordinary* wave, a description which is quite appropriate as we shall see from its properties. If $k$ is at an angle $\theta$ with respect to the axis of symmetry the phase velocity $v_e$ (defined as $\omega/|k|$) of this wave is given by

$$v_e^2 = \frac{\cos^2\theta}{\mu\varepsilon_{\parallel}} + \frac{\sin^2\theta}{\mu\varepsilon_{\perp}}. \qquad (2.55)$$

The phase velocity is the same as that for the ordinary wave in the direction of the axis of symmetry only. The surfaces of constant phase are planes perpendicular to $k$ which move in the direction of $k$ with a speed $v_e$.

The direction of energy propagation is that of the vector $E \wedge H$

which must be also that of the group velocity $v_g$. If the group velocity is calculated from (2.34) using the definition (2.34), we find

$$v_g = \frac{\omega}{\mu\varepsilon_\parallel}\left(k_1,\ k_2,\ \frac{\varepsilon_\parallel}{\varepsilon_\perp}k_3\right), \tag{2.56}$$

where the $\omega$-dependence of $\varepsilon_\parallel$, $\varepsilon_\perp$ and $\mu$ has been neglected. It is clear from (2.56) that the energy is propagated at an angle to $k$ in a direction which lies in the principal section. Hence, $E \wedge H$ must also lie in the principal section. From (2.48) we see that $H$ is perpendicular to $k$ so that $D$ and $E$ must both lie in the principal section but in different directions (see Fig. 4). Therefore the extraordinary wave is

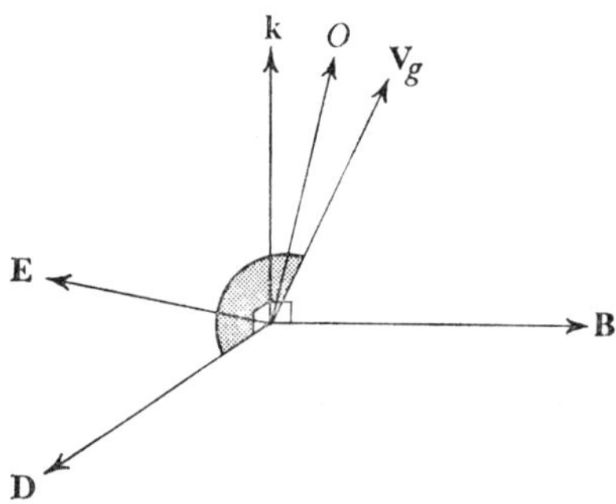


Fig. 4 The vectors associated with the extraordinary wave in a uniaxial crystal. $O$ is the symmetry axis of the crystal. D, E, k, 0 and $V_g$ all lie in the same plane

plane polarized at right angles to the ordinary wave. This result could have been deduced on more general grounds because equation (2.51) could have been expressed as an eigenvalue problem with equation (2.52) determining the eigenvalues. The two eigenvalues which correspond to the phase velocities of the two waves are distinct so the corresponding field vectors must be orthogonal.

In biaxial crystals there are two directions in which the two types of wave which can be propagated have the same phase velocity. These directions bear no simple relation to the crystallographic axes. The propagation of the waves in this case is rather more complicated and will not be dealt with here.

## 2.7 Spherical waves

So far we have looked at only plane wave solutions of Maxwell's equations. This is not such a limitation as at first sight it might appear, because more general solutions can be constructed from linear combinations of plane waves with different $k$ values. For plane waves with a continuous spectrum of $k$ values and amplitudes $E_0(k)$,

$$E(r_1 t) = \frac{1}{\sqrt{8\pi^3}} \iiint_{-\infty}^{\infty} E_0(k)e^{i(\omega_k t - k \cdot r)} dk_1 dk_2 dk_3 \qquad (2.57)$$

is a solution of the wave equation. The relation between $\omega_k$ and $k$ depends upon the propagating medium; in free space it is $\omega_k = ck$.

Any spherical or axial symmetry can be exploited by generating solutions of the equations in the appropriate coordinate system. In spherical coordinates $(r, \theta, \phi)$ a whole class of product solutions can be found by the method of separation of variables. The vectors $E$ and $H$ are written in terms of their components $(E_r, E_\theta, E_\phi)$, $(H_r, H_\theta, H_\phi)$ with respect to the spherical coordinate system. Unfortunately, (2.28) does not separate into independent equations for each component as in the case of the cartesian coordinate system. This would involve the manipulation of several simultaneous equations which can be avoided by introducing a suitable scalar potential function $\psi_\omega$, related to $E$ and $H$ by the equations

$$E_\omega = r \wedge \operatorname{grad} \psi_\omega, \qquad (2.58)$$

and

$$H_\omega = \frac{i}{\mu\omega} \operatorname{curl} (r \wedge \operatorname{grad} \psi_\omega). \qquad (2.59)$$

These fields satisfy Maxwell's equations if $\psi_\omega$ satisfies the scalar wave equation

$$\nabla^2 \psi_\omega + k^2 \psi_\omega = 0, \qquad (2.60)$$

where $k^2 = \mu\varepsilon\omega^2$.

For a product solution $\psi\omega = R(r)\Theta(\theta)\Phi(\phi)$, (2.60) can be separated into three ordinary differential equations

$$\frac{d^2\Phi}{d\phi^2} + M\Phi = 0, \qquad (2.61)$$

$$\frac{1}{\sin\theta} \frac{d}{d\theta}\left(\sin\theta \frac{d\Theta}{d\theta}\right) + \left[L - \frac{M}{\sin^2\theta}\right]\Theta = 0, \qquad (2.62)$$

$$\frac{d}{dr}\left(r^2 \frac{dR}{dr}\right) - [L - k^2 r^2]R = 0. \qquad (2.63)$$

It is convenient to write $M = m^2$ so that the solution of (2.61) can be written as

$$\Phi = e^{\pm im\phi}. \tag{2.64}$$

Equation (2.62) can be transformed into the standard form of Legendre's equation in terms of the variable $\cos\theta$. To use the standard notation $L = l(l+1)$, the solutions are

$$\Theta_{lm}(\theta) = \begin{cases} P_l^m(\cos\theta) \\ Q_l^m(\cos\theta) \end{cases}, \tag{2.65}$$

where $P_l^m$ and $Q_l^m$ are associated Legendre functions (see Appendix B).

The third equation (2.63) can be written in a familiar form in terms of $Z$ and $\eta$, where $R = \eta^{-\frac{1}{2}}z$ and $\eta = kr$,

$$\frac{d^2z}{d\eta^2} + \frac{1}{\eta}\frac{dz}{d\eta} - \left[\frac{(l+\frac{1}{2})^2}{\eta^2} - 1\right]z = 0, \tag{2.66}$$

which is Bessel's equation of order $(l+\frac{1}{2})$. The solutions are Bessel functions, $J_{l+\frac{1}{2}}(kr)$ and $Y_{l+\frac{1}{2}}(kr)$ of the first and second kind (see Appendix B). The complete solution for $\psi_\omega$ is a linear combination of products of the form

$$\psi_{lm}(\omega) = \begin{cases} j_l(kr)P_l^m(\cos\theta) \\ y_l(kr)Q_l^m(\cos\theta) \end{cases} e^{\pm im\phi}, \tag{2.67}$$

where

$$j_e = \sqrt{\frac{\pi}{2kr}}\,J_{l+\frac{1}{2}}(kr), \quad y_e = \sqrt{\frac{\pi}{2kr}}\,Y_{l+\frac{1}{2}}(kr). \tag{2.68}$$

The functions $j_l(kr)$ and $y_l(kr)$ are known as the spherical Bessel functions of order $l$. The combination $j_l(kr) - iy_l(kr)$ behaves asymptotically like an outgoing wave for large $r$ when combined with the time dependence of the wave $e^{i\omega t}$,

$$j_l(kr) - iy_l(kr) \sim \frac{e^{-ikr + \frac{i\pi}{2}(l+1)}}{kr}. \tag{2.69}$$

Similarly, $j_l(kr) + iy_l(kr)$, when combined with the factor $e^{i\omega t}$ leads to a spherical incoming wave; its asymptotic form can be found by replacing $i$ by $-i$ in (2.69).

In general the solutions represent a combination of waves propagated either outwards or inwards along the radius vector $r$. The wave form found from (2.58) (2.59) is known as a *transverse electric* (T.E.) wave because its electric vector is always perpendicular to $r$. An independent *transverse magnetic* (T.M.) wave can be found from the same potential by interchanging the roles of $E_\omega$ and $H_\omega$ such that

$$H_\omega = r \wedge \operatorname{grad} \psi_\omega, \tag{2.70}$$

$$E_\omega = -\frac{i}{\varepsilon\omega} \operatorname{curl} (r \wedge \operatorname{grad} \psi_\omega). \tag{2.71}$$

We shall find later that these types of waves are generated by small oscillating electric and magnetic dipoles.

## PROBLEMS 2

1. Show that the general plane wave,
$$E = E_0 F(\alpha t + a \cdot r), \quad H = H_0 F(\alpha t + a \cdot r),$$
where $F$ is an arbitrary function with continuous derivatives, satisfies Maxwell's equations in free space if (a) $E_0$, $H_0$ and $a$ form an orthogonal triad, (b) $\sqrt{\varepsilon_0}\,|E_0| = \sqrt{\mu_0}\,|H_0|$ and (c) $|a|^2 = \alpha^2 \mu_0 \varepsilon_0$.

2. Find the Fourier transform of the electric field $E(t)$ for the pulse
$$E(t) = E_0 e^{-i\omega_0 t - \Gamma|t|}$$
for all $t$, and for the wave of finite duration,

$$E(t) = \begin{cases} E_0 e^{-i\omega_0 t} & -\dfrac{1}{\Gamma} \leqq t \leqq \dfrac{1}{\Gamma} \\[2ex] 0 & -\dfrac{1}{\Gamma} > t > \dfrac{1}{\Gamma} \end{cases}.$$

Show that they are sharply peaked at $\omega = \omega_0$ for $\Gamma \ll 1$ with a width of $2\Gamma$.

3. Consider an electron of charge $e$ and mass $m$ whose $x$-coordinate $x_j$ satisfies the equation
$$\frac{d^2 x_j}{dt^2} + \gamma_j \frac{dx_j}{dt} + \omega_j^2 x_j = \frac{eE_0}{m} e^{i\omega t}$$
in the presence of an electric field $E_0 e^{i\omega t}$. If there are $n_j$ such electrons $(j = 1, \ldots n_0)$ per unit volume, show that the polarization $P(= \sum_j n_j x_j e)$ induced in the $x$-direction by the field is given by

$$= E_0 e^{i\omega t} \sum_{j=1}^{n_0} \frac{n_j e^2 / m}{\omega_j^2 - \omega^2 + i\omega\gamma_j}.$$

Hence find the real and imaginary parts of the dielectric constant, and examine their variation as a function of $\omega$ about the point $\omega = \omega_j$.

# CHAPTER 3

# Reflection, Refraction and Propagation in Wave Guides

## 3.1 Dielectric-dielectric boundary

The wave solutions of Maxwell's equations considered in the last chapter correspond to waves travelling in an unbounded medium. In a bounded medium, the fields must satisfy conditions (1.42), (1.45), (1.47) and (1.49) at the boundaries, and this may place considerable restrictions on the type of waves that can exist within the medium.

Before considering waves in a medium which is completely enclosed, we shall consider the way in which a wave is affected by a single boundary between two semi-infinite homogeneous media. Let the boundary be the plane $z = 0$, let medium $1(z<0)$ be a dielectric characterized by constants $\varepsilon_1$, $\mu_1$ and let medium $2(z>0)$ be a dielectric with constants $\varepsilon_2$, $\mu_2$. Any wave incident on this interface will, in general, be partially reflected and partially transmitted. For a plane wave with fields $(E_1, H_1)$ incident at the interface from medium 1, let the corresponding reflected wave have fields $(E_2, H_2)$ and the transmitted wave have fields $(E_3, H_3)$. As these are plane waves

$$E_\alpha = E_{0\alpha} e^{i(\omega_\alpha t - k_\alpha \cdot r)},$$

$$B_\alpha = \frac{k_\alpha \wedge E_{0\alpha}}{\omega_\alpha} e^{i(\omega_\alpha t - k_\alpha \cdot r)}, \quad \alpha = 1, 2, 3, \tag{3.1}$$

where $k_\alpha$ is the propagation vector associated with each wave,

$$\left|\frac{k_1}{\omega_1}\right| = \left|\frac{k_2}{\omega_2}\right| = \sqrt{\varepsilon_1 \mu_1}, \quad \left|\frac{k_3}{\omega_3}\right| = \sqrt{\mu_2 \varepsilon_2}. \tag{3.2}$$

The conditions imposed on the fields at the boundary $z = 0$ apply for all $t$ and at all points over the plane so the phase factors $(\omega_\alpha t - k_\alpha \cdot r)$ of all three waves must be equal. This means that the frequency cannot be changed either on reflection or transmission. Hence $\omega_1 = \omega_2 = \omega_3$. The condition $k_1 \cdot r = k_2 \cdot r = k_3 \cdot r$ at $z = 0$ means that the projections of the propagation vectors on the plane

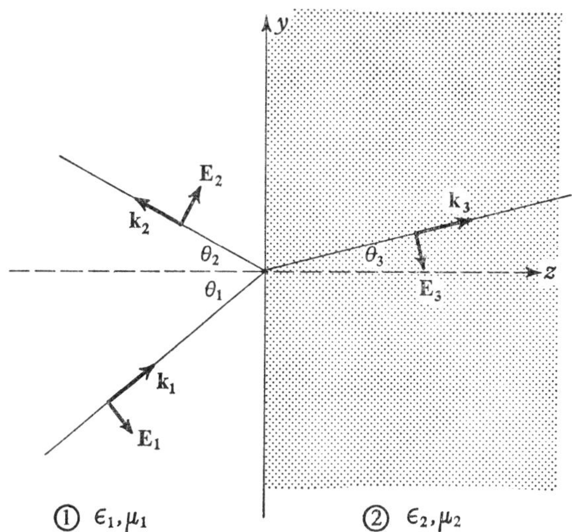


Fig. 5 Reflection and refraction of a plane wave which is
polarized with the electric vector in the plane of incidence

$z = 0$ must be equal. If the incident wave lies in the $yz-$ plane and
the propagation vectors $k_1$, $k_2$, $k_3$ make angles $\theta_1$, $\pi-\theta_2$, $\theta_3$ with
$z$-axis respectively (see Fig. 5), then $k_{1y} = k_{2y}$ implies $\theta_1 = \theta_2$ and
$k_{1y} = k_{3y}$ implies

$$\sqrt{\varepsilon_1\mu_1}\,\sin\theta_1 = \sqrt{\varepsilon_2\mu_2}\,\sin\theta_3, \tag{3.3}$$

or
$$\frac{\sin\theta_1}{\sin\theta_3} = n, \tag{3.4}$$

where $n\,(=\sqrt{\varepsilon_2\mu_2/\varepsilon_1\mu_1})$ can be identified as the refractive index of
medium 2 relative to medium 1 because (3.4) expresses the well-
known Snell's law of optical refraction.

The amplitudes of the reflected and refracted waves depend upon
the direction of polarization of the incident plane wave. For a wave
with its magnetic vector in the plane of incidence the amplitudes of
the reflected and refracted waves are related to those of the incident
wave by

$$\frac{E_2}{E_1} = \frac{\sin(\theta'-\theta)}{\sin(\theta'+\theta)}, \qquad \frac{H_2}{H_1} = \frac{\sin(\theta'-\theta)}{\sin(\theta'+\theta)}, \tag{3.5}$$

$$\frac{E_3}{E_1} = \frac{2\sin\theta'\cos\theta}{\sin(\theta'+\theta)}, \qquad \frac{H_3}{H_1} = \frac{\sin 2\theta}{\sin(\theta+\theta_t')}. \tag{3.6}$$

These expressions can be deduced from (3.1) and (3.2) by using the continuity conditions on the tangential components of $E$ and $H$ at the interface. The notation has been simplified to $\theta = \theta_1 = \theta_2$ and $\theta' = \theta_3$.

The rate at which the energy of the incident wave falls upon a unit area of the interface is equal to the projection of the Poynting vector along the normal $E_1 H_1 \cos \theta$. The corresponding rates for the transmitted and reflected waves are $E_3 H_3 \cos \theta'$ and $E_2 H_2 \cos \theta$; the ratio of these to the incident rate defines the transmission and reflection coefficients $T$ and $R$ respectively. From (3.5) and (3.6) we find that

$$T = \frac{\sin 2\theta \sin 2\theta'}{\sin^2 (\theta + \theta')} \quad \text{and} \quad R = \frac{\sin^2 (\theta' - \theta)}{\sin^2 (\theta' + \theta)}, \tag{3.7}$$

which are in agreement with the conservation of energy requirement that $R + T = 1$.

The amplitudes of the reflected and transmitted waves for an incident wave which has its electric vector in the plane of incidence are

$$\frac{E_2}{E_1} = \frac{\tan (\theta - \theta')}{\tan (\theta + \theta')}, \qquad \frac{H_2}{H_1} = \frac{\tan (\theta - \theta')}{\tan (\theta + \theta')}, \tag{3.8}$$

$$\frac{E_3}{E_1} = \frac{2 \sin \theta \cos \theta}{\sin (\theta + \theta') \cos (\theta - \theta')}, \qquad \frac{H_3}{H_1} = \frac{\sin 2\theta}{\sin (\theta + \theta') \cos (\theta - \theta')}. \tag{3.9}$$

A feature of these results which deserves special comment is the absence of a reflected wave for $\theta + \theta' = \pi/2$ when the incident wave is polarized with its electric vector in the plane of incidence. The polarization of light, and this particular phenomenon, were known in the early nineteenth century well before the electromagnetic nature of light was discovered. The relation between the refractive index and the angle of incidence at which this effect occurs is $\tan \theta = n$, which is generally referred to as Brewster's law. Some insight into the reasons for its occurrence can be gained by examining the way in which the reflected and refracted waves are generated. When the electromagnetic wave passes into medium 2 the electrons of the medium are set into forced oscillations by the electric and magnetic fields of the wave, an oscillating dipole moment is induced which generates secondary waves. In Chapter 5, where we consider sources of electromagnetic waves, we shall find that an oscillating electric dipole emits waves in all directions except along its axis. The axis of the induced dipole is along the electric vector of the wave in medium 2, so when $\theta + \theta' = \pi/2$, the direction of reflection, $\pi - \theta$, lies along this axis, but there is no reflected wave because there is no source of secondary waves propagated in this direction. The phenomenon does not occur for a wave with its electric vector perpendicular to the plane of incidence because the dipole that this induces radiates secondary waves in all directions in the plane of incidence.

When $n < 1$ the light travels from an optically denser to an optically rarer medium and equations (3.5) to (3.9) are valid only if $\theta < \sin^{-1} n$. As $\theta$ approaches the limiting value $\sin^{-1} n$ the angle the refracted wave makes with the $z$-axis approaches grazing angles $\theta' \sim \pi/2$. When $\theta$ exceeds $\sin^{-1} n$ the incident wave is completely reflected. This is the familiar optical phenomenon of total internal reflection. To describe it in detail we must take the complex solution of (3.4) for $\theta_3$.

If we consider a wave polarized perpendicularly to the plane of incidence and re-write (3.5) for the reflected electric vector in terms of $\theta$ and $n$, we find

$$\frac{E_2}{E_1} = \frac{\cos\theta - i\sqrt{\sin^2\theta - n^2}}{\cos\theta + i\sqrt{\sin^2\theta - n^2}}. \tag{3.10}$$

For $\sin\theta > n$, the reflected and incident waves differ by a phase factor only so that

$$\left|\frac{E_2}{E_1}\right|^2 = 1. \tag{3.11}$$

This proves that the wave is completely reflected. A similar argument can be used to deduce the same result for a wave polarized within the plane of incidence.

The difference in phase between the incident and reflected waves is due to the penetration of the wave from medium 1 into medium 2 before it is eventually completely reflected. The wave that exists in medium 2 is strongly attenuated but is not absorbed. To calculate the solution in this region let us consider a wave with its electric vector perpendicular to the plane of incidence

$$E_{3x} = E_{0x}e^{i(\omega t - k_3 y \sin\theta_3 - k_3 z \cos\theta_3)}. \tag{3.12}$$

The solution of (3.4) for $\theta_3$ must be of the form $\theta = \pi/2 + i\alpha$, where $\alpha$ is real and has two possible values, one positive, the other negative. Substituting this form for $\theta_3$ into (3.12) we find

$$E_{3x} = E_{0x}e^{i(\omega t - k_3 y \cosh\alpha) - k_3 z \sinh\alpha}, \tag{3.13}$$

where $\alpha$ must be taken as positive because a negative value would give a wave which exponentially increased as $z \to \infty$. The magnetic vector associated with this wave can be calculated from (2.4) to give

$$H_{3y} = -iE_{0x}\sqrt{\frac{\varepsilon_2}{\mu_2}}\sinh \alpha e^{i(\omega t - k_3 y \cos \alpha) - k_3 z \sinh \alpha}, \quad (3.14)$$

$$H_{3z} = -E_{0x}\sqrt{\frac{\varepsilon_2}{\mu_2}}\cosh \alpha e^{i(\omega t - k_3 y \cos \alpha) - k_3 z \sinh \alpha}. \quad (3.15)$$

This is not a plane wave because the planes of constant phase are perpendicular to the $y$-direction, whereas the planes of constant amplitude are perpendicular to the $z$-direction. There is a component of the Poynting vector $E_3 \wedge H_3$ normal to the interface which changes sign periodically and its time average $\bar{S}_z$, which is equal to $\mathrm{Re}\,\dfrac{E_x H_y^*}{2}$ (see Problem 1, Chapter 1), is zero. There is also a component of the Poynting vector is the $y$-direction, its time average $\bar{S}_y$ is non-zero and is given by

$$S_y = \tfrac{1}{2}\sqrt{\frac{\varepsilon_2}{\mu_2}}\,|E_{0x}|^2 \cosh \alpha e^{-2k_3 z \sinh \alpha}. \quad (3.16)$$

This implies that there is a net flow of energy in medium 2 parallel to the interface. If we could follow the progress of a point on the wave front as it passed into medium 2, we would observe it being displaced in the positive $y$ and $z$ direction before its motion in the $z$-direction reversed and it re-entered medium 1 further along the interface. As a plane wave has an infinite wave front the displacement of the wave in the $y$-direction makes no essential difference. The existence of this type of wave has been confirmed experimentally and its displacement has been detected by using a beam of finite cross-section (see [2]).

By the use of a planar dielectric slab or a series of planar slabs, a composite boundary medium can be constructed with special transmission and reflection properties. Matrix methods enable the solution of the general problem to be constructed from a knowledge of the fields in one particular slab. Consider a series of parallel planes $z = z_s$ where $s = 1, \ldots, n-1$, and let the slabs between the planes $z = z_{s-1}$ and $z = z_s$ be a dielectric with constants $\varepsilon_s$, $\mu_s$. For a wave polarized perpendicularly to the plane of incidence, the fields will be a sum of reflected and transmitted waves of the form

$$E_x = \{A_s e^{-ik_s(z-z_s)\cos\theta_s} + B_s e^{ik_s(z-z_s)\cos\theta_s}\} e^{i(\omega t - k_s y \sin\theta_s)}, \qquad (3.17)$$

$$H_y = \sqrt{\frac{\varepsilon_s}{\mu_s}} \cos\theta_s \{A_s e^{-ik_s(z-z_s)\cos\theta_s} + B_s e^{ik_s(z-z_s)\cos\theta_s}\} e^{i(\omega t - k_s y \sin\theta_s)},$$

$$(3.18)$$

where $k_s = \omega\sqrt{\mu_s \varepsilon_s}$ and $\theta_s$ is the angle the transmitted wave makes with the $z$-axis. If the field values on the plane $z = z_s$ at the point $x = 0$, $y = 0$ and $t = 0$, are denoted by $E_s$ and $H_s$, then

$$E_s = A_s + B_s, \qquad (3.19)$$

and

$$H_s = \beta_s(A_s - B_s), \qquad (3.20)$$

where

$$\beta_s = \sqrt{\frac{\varepsilon_s}{\mu_s}} \cos\theta_s.$$

From the continuity conditions on the tangential components of $E$ and $H$ at $z = z_{s-1}$, we obtain the relations

$$E_{s-1} = \cos\alpha_s E_s + i/\beta_s \sin\alpha_s H_s, \qquad (3.21)$$

and

$$H_{s-1} = \cos\alpha_s H_s + i\beta_s \sin\alpha_s E_s, \qquad (3.22)$$

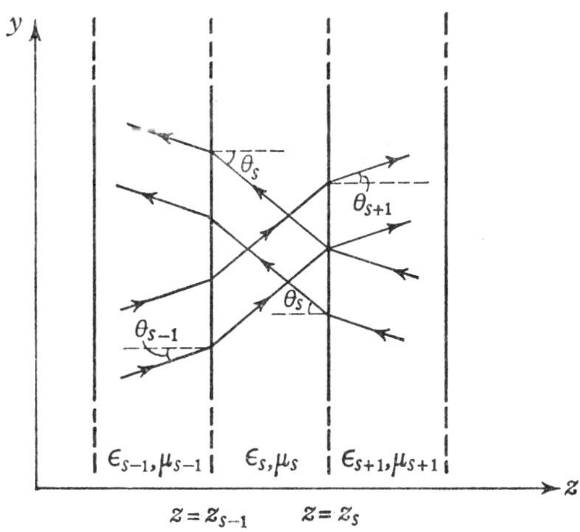


Fig. 6 The transmission and reflection of a plane wave through a series of slabs.

which can be written in the form

$$\begin{pmatrix} E_{s-1} \\ H_{s-1} \end{pmatrix} = K_s \begin{pmatrix} E_s \\ H_s \end{pmatrix}, \tag{3.23}$$

where $K_s$ is the matrix,

$$\begin{pmatrix} \cos \alpha_s & i/\beta_s \sin \alpha_s \\ i\beta_s \sin \alpha_s & \cos \alpha_s \end{pmatrix}, \tag{3.24}$$

with $\alpha_s = k_s(z_s - z_{s-1}) \cos \theta_s$.

The fields on the first and last planes can be related by taking the product of the appropriate matrix for each slab to give

$$\begin{pmatrix} E_1 \\ H_1 \end{pmatrix} = K_2 K_3 ... K_{n-1} \begin{pmatrix} E_{n-1} \\ H_{n-1} \end{pmatrix}, \tag{3.25}$$

or

$$\begin{pmatrix} E_1 \\ H_1 \end{pmatrix} = \begin{pmatrix} K_{11} & K_{12} \\ K_{21} & K_{22} \end{pmatrix} \begin{pmatrix} E_{n-1} \\ H_{n-1} \end{pmatrix}, \tag{3.26}$$

where $K$ is the product of the matrices in (3.25).

The usual situation one has to consider is a plane wave of known amplitude $E_0$ arriving at $z = z_1$, resulting in a reflected wave of amplitude $E_R$, and a transmitted wave of amplitude $E_T$ in the dielectric $n$ beyond the plane $z = z_{n-1}$. These amplitudes are related to the fields on the first and last planes by

$$\begin{cases} E_1 = E_0 + E_R, \\ H_1 = \beta_1(E_0 - E_R), \end{cases} \begin{cases} E_{n-1} = E_T, \\ H_{n-1} = \beta_n E_T. \end{cases} \tag{3.27}$$

From (3.26) and (3.27) the ratio of the amplitude of the reflected to incident waves is

$$\frac{E_R}{E_0} = \frac{\beta_1(K_{11} - K_{12}\beta_n) - (K_{21} + K_{22}\beta_n)}{(K_{11} + K_{12}\beta_n)\beta_1 + (K_{21} + K_{22}\beta_n)} \tag{3.28}$$

and the ratio of the transmitted to incident amplitudes is

$$\frac{E_T}{E_0} = \frac{2\beta_1}{(K_{11} + K_{12}\beta_n)\beta_1 + (K_{21} + K_{22}\beta_n)}. \tag{3.29}$$

For a single slab of dielectric $\varepsilon_2$, $\mu_2$ of thickness $a$ dividing a dielectric $\varepsilon_1$, $\mu_1$, from a dielectric $\varepsilon_3$, $\mu_3$, we can use (3.28) and (3.29) with $n = 3$, and for $K$ the single matrix $K_2$, i.e.

$$\frac{E_R}{E_0} = \frac{\beta_2(\beta_1 - \beta_3) \cos \alpha + i \sin \alpha(\beta_1\beta_3 - \beta_2^2)}{\beta_2(\beta_1 + \beta_3) \cos \alpha + i \sin \alpha(\beta_1\beta_3 + \beta_2^2)}, \tag{3.30}$$

$$\frac{E_T}{E_0} = \frac{2\beta_1\beta_1}{\beta_2(\beta_1+\beta_3)\cos\alpha + i\sin\alpha(\beta_1\beta_3+\beta_2^2)}, \tag{3.31}$$

where $\alpha = \omega a \cos\theta_2$, and $\theta_2$ is the angle of propagation in dielectric 2. The corresponding reflection and transmission coefficients are

$$R = \frac{\beta_2^2(\beta_1-\beta_3)^2\cos^2\alpha + (\beta_1\beta_3-\beta_2^2)^2\sin^2\alpha}{\beta_2^2(\beta_1+\beta_3)^2\cos^2\alpha + (\beta_1\beta_3+\beta_2^2)^2\sin^2\alpha}, \tag{3.32}$$

and

$$T = \frac{4\beta_1\beta_3\beta_2^2}{\beta_2^2(\beta_1+\beta_3)^2\cos^2\alpha + (\beta_1\beta_3+\beta_2^2)^2\sin^2\alpha}. \tag{3.33}$$

The reflection coefficient is a maximum or minimum as a function of the width of the slab when $\sin 2\alpha = 0$ or $\alpha = \dfrac{m\pi}{2}$, where $m$ is an integer.

For odd values of $m$

$$R = \frac{(\beta_1\beta_3-\beta_2^2)^2}{(\beta_1\beta_3+\beta_2^2)^2}. \tag{3.34}$$

There is no reflected wave when $\beta_1\beta_3 = \beta_2^2$, a result which has been applied in the design of certain optical systems where it is desirable to keep reflections to a minimum. The surfaces are covered with a film of dielectric which has a dielectric constant such that this equality is approximately satisfied at the normal angle of incidence of the light. For even values of $m$

$$R = \left(\frac{\beta_1-\beta_3}{\beta_1+\beta_3}\right)^2, \tag{3.35}$$

which is independent of medium 2 and is the reflection coefficient of a single interface between dielectrics 1 and 3. When dielectric 3 is the same as dielectric 1, there is no reflected wave.

Equations (3.34) (3.35) can form the basis of an explanation of the colour of thin films, such as oil on water, when illuminated by an extended source of white light. For a film of given thickness the reflected light will be of maximum or minimum intensity if

$$\cos\theta_2 = \frac{m\lambda}{2a} \tag{3.36}$$

for light of wavelength $\lambda$. This is due to interference between light reflected at the first and second planes, a maximum occurs when these

D

are in phase, a minimum when they are out of phase. The angles at which the intensities of the reflected wave are maxima depends upon the wavelength, so when viewed at a given angle, the colour of the component of white light that satisfies the condition for constructive interference is dominant.

Though we have deduced some of the well-known laws of optics by considering the behaviour of plane waves, the conditions under which they are strictly valid do not correspond to any real physical situation. A plane wave has a constant amplitude over infinite planes and exists for an infinite time, whereas optical experiments are conducted with light passing through finite apertures for a finite duration. The effect of an aperture is to cause the electromagnetic waves to be diffracted and the intensity of the wave to be modulated in a complicated way which we shall discuss in Chapter 6. If the wavelength of the wave is extremely small compared with the dimensions of the aperture, the diffraction effects are negligible and light is either cut off by the aperture or passes through it unaffected so that a well-defined beam and shadow are formed.

Geometrical optics is the theory of light waves when diffraction effects can be ignored. It is a small wavelength or high frequency approximation based on the assumption that waves behave locally as plane waves.

If electric and magnetic field vectors

$$E = E_0 e^{i\phi}, \quad H = H_0 e^{i\phi}, \tag{3.37}$$

are to describe local plane waves of frequency $\omega$, then $\phi$, which is known as the *eikonal*, must be of the form $\phi = k\psi - \omega t$, where $k$ is $\omega\sqrt{\varepsilon_0\mu_0}$, and $\psi$, $E_0$ and $H_0$ are slowly varying functions of the spatial coordinates. This approximation cannot be valid in certain regions such as those near the edge of a shadow where the amplitudes $E_0$, $H_0$ must vary rapidly. If we substitute (3.37) into Maxwell's equations and retain terms to zero order in $1/\omega$ then

$$E_0 \cdot \nabla\psi = 0, \qquad\qquad H_0 \cdot \nabla\psi = 0, \tag{3.38}$$

$$\omega\mu H_0 = -kE_0 \wedge \operatorname{grad}\psi, \qquad \omega\varepsilon E_0 = kH_0 \wedge \operatorname{grad}\psi. \tag{3.39}$$

From (3.39) we can deduce an equation for $\psi$,

$$(\nabla\psi)^2 = n^2, \tag{3.40}$$

where $n$ is $(\varepsilon\mu/\varepsilon_0\mu_0)^{\frac{1}{2}}$, the refractive index of the medium. To zero order in $\omega$ this equation is valid for an inhomogeneous dielectric in which the refractive index varies slowly over distances large compared to the wavelength. The surfaces of given phase are known as the *wave fronts*. They can be divided up into a large number of small areas that are approximately planar; the locus of one of these areas as the wave travels generates a *ray*. The energy of a local plane wave is propagated along the direction of the local Poynting vector $E_0 \wedge H_0$ which is in the direction of grad $\psi$. It is also the direction normal to the wave front and, therefore, along the direction of the local ray. If $r(s)$ is the position vector locating a point $P$ on the ray, where $s$ is the path length to $P$, then $\dfrac{dr}{ds}$ is a unit vector in the direction of the ray at the point $P$. The ray path must, therefore, satisfy the equation

$$\frac{dr}{ds} = \frac{\operatorname{grad} \psi}{n}. \tag{3.41}$$

From (3.40) and (3.41) we can eliminate the variable $\psi$. If we differentiate (3.41) with respect to $s$, we have

$$\frac{d}{ds}\left(n\,\frac{dr}{ds}\right) = \frac{d}{ds}(\operatorname{grad}\psi), \tag{3.42}$$

which can be rewritten in the form

$$\frac{d}{ds}\left(n\,\frac{dr}{ds}\right) = \left(\frac{dr}{ds}\cdot\operatorname{grad}\right)\operatorname{grad}\psi, \tag{3.43}$$

(see Appendix B).

On substituting for $\dfrac{dr}{ds}$ from (3.41) we find

$$\frac{d}{ds}\left(n\,\frac{dr}{ds}\right) = \left(\frac{\operatorname{grad}\psi\cdot\operatorname{grad}}{n}\right)\operatorname{grad}\psi. \tag{3.44}$$

It is possible to rewrite this as

$$\frac{d}{ds}\left(n\,\frac{dr}{ds}\right) = \frac{1}{2n}\operatorname{grad}(\operatorname{grad}\psi)^2, \tag{3.45}$$

so that $(\nabla\psi)^2$ can be eliminated by making a substitution from (3.40). We then obtain the equation for the ray path in terms of the refractive index

$$\frac{d}{ds}\left(n\,\frac{dr}{ds}\right) = \operatorname{grad} n. \tag{3.46}$$

If we are given the functional dependence of $n$ on the spatial coordinates, this equation can be used to calculate the ray paths. In the simplest case of a homogeneous medium $n = $ constant, the solutions are straight lines $r = sa + b$, where $a$ and $b$ are constant vectors.

Now $\dfrac{dr}{ds}$ is a unit vector along the ray, therefore its derivative $\dfrac{d^2r}{ds^2}$ is a vector normal to the ray path at the point $r(s)$, and its modulus is the curvature $1/\rho$ at the same point. (3.44) can be written in the form

$$\frac{dn}{ds}\frac{dr}{ds} + n\frac{d^2r}{ds^2} = \text{grad } n. \tag{3.47}$$

If $m$ is a unit vector along the normal at the point $r(s)$, the scalar product of (3.47) with $m$ gives the equation

$$\frac{1}{\rho} = m \cdot \frac{\text{grad } n}{n}. \tag{3.48}$$

This equation clearly indicates that rays curve in a direction of increasing refractive index and away from the direction of decreasing refractive index.

In general the problem of describing the propagation of electromagnetic waves in a dielectric in which $\varepsilon$ and $\mu$ vary over macroscopic distances is very difficult. For a few very special forms of $n(x, y, z)$ exact solutions of Maxwell's equations can be found. In most cases an approximate solution is the most one can hope to obtain. One type of approximation is to take a simple model of the problem which can be solved exactly. For instance, if the variations in $n(x, y, z)$ are in the $z$-direction only, the original inhomogeneous dielectric is replaced by a series of dielectric slabs such that the dielectric constant is constant within each slab but the variation from slab to slab approximates to the variation in the original inhomogeneous medium. The propagation within the simplified model can be solved exactly by the techniques outlined in (3.18) to (3.29). The solution should approximate to that in the continuous medium as the number of slabs is increased and their width decreased. Other methods depend upon making a mathematical approximation which is valid to a reasonable accuracy for a certain range of parameters. Geometrical optics is typical of this type of approach; the accuracy in this case depends upon the wavelength of the waves and is asymptotically correct as this tends to zero.

## 3.2 Dielectric-conductor boundary

Let us reconsider the situation portrayed in Fig. 5, but with medium 2 having a finite conductivity $\sigma_2$. Reflection and refraction occur as for a dielectric-dielectric boundary but the transmitted wave must be attenuated and its electric vector must be of the form

$$E = E_0 e^{-k'' \cdot r} e^{i(\omega t - k' \cdot r)}. \tag{3.49}$$

The direction of $k''$ is determined by the condition that the amplitude must be constant over the plane $z = 0$. This means that $k$ must be perpendicular to the plane, and therefore in the $z$-direction. The relation between the angle of incidence $\theta$ and the angle of refraction $\theta'$ can be obtained by equating the phases of the incident and refracted waves on $z = 0$

$$k' \sin \theta' = \omega \sqrt{\varepsilon_1 \mu_1} \sin \theta. \tag{3.50}$$

Given that the angle between the vector $k'$ and $k''$ is $\theta'$, (2.45) can be used to calculate their magnitudes. If medium 2 is a very good conductor such that $\varepsilon_2 \omega / \sigma_2 \ll 1$, the first term on the right-hand side of (2.45) can be neglected in comparison with the second, which leads to the solution

$$k' = k'', \qquad 2k'^2 \cos \theta' = \omega \sigma_2 \mu_2. \tag{3.51}$$

Substituting this expression for $k'$ into (3.50) gives a relation between $\theta$ and $\theta'$, which is the dielectric-conductor equivalent of Snell's Law

$$\sin \theta' \tan \theta' \simeq \frac{2\omega \varepsilon_1 \mu_1}{\sigma_2 \mu_2} \sin^2 \theta. \tag{3.52}$$

The planes perpendicular to $\theta'$ are the planes of constant phase for the transmitted wave. In the approximation $\omega \varepsilon_1 / \sigma_2 \ll 1$ for a good conductor, $\theta'$ will be very small whatever the angle of incidence $\theta$. We shall calculate the fields only in the special case of normal incidence. Using the continuity conditions on the tangential component of $E$ and the normal component of $D$, we find

$$\frac{E_2}{E_1} = \frac{1 - \dfrac{\gamma}{\mu_2 \omega} \sqrt{\dfrac{\mu_1}{\varepsilon_1}}}{1 + \dfrac{\gamma}{\mu_2 \omega} \sqrt{\dfrac{\mu_1}{\varepsilon_1}}}, \quad \frac{E_3}{E_1} = \frac{2}{1 + \dfrac{\gamma}{\mu_2 \omega} \sqrt{\dfrac{\mu_1}{\varepsilon_1}}}, \tag{3.53}$$

where $\gamma = |k'| + i|k''|$.

In a good conductor the transmitted wave is strongly attenuated. The proportion of energy reflected is measured by the reflection coefficient, which to lowest order in $\omega_1 \varepsilon_1 / \sigma_2$ is

$$R \simeq 1 - 2 \sqrt{2 \left( \frac{\mu_2}{\mu_1} \right) \left( \frac{\omega \varepsilon_1}{\sigma_2} \right)}. \qquad (3.54)$$

The difference between the reflection coefficient and unity is a measure of the intensity of the transmitted wave. Eventually all the transmitted energy must be absorbed in the dissipative processes

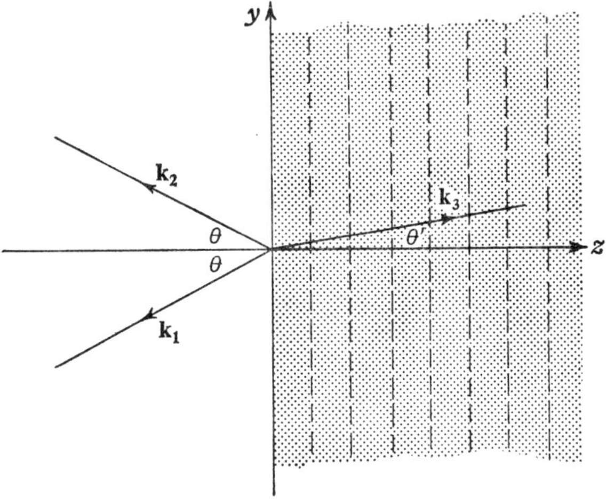


Fig. 7 Reflection and refraction of a plane wave at a dielectric-conductor interface – – – indicate planes of constant amplitude

within the conductor because the amplitude of the transmitted wave drops to zero as $z \to \infty$. In the limit of a perfect conductor $\sigma_2 \to \infty$ the incident wave is completely reflected; because there is no penetration of the wave into the conductor there is no absorption, and therefore no energy loss.

If we introduce a second dielectric-conductor interface so that medium 1 is sandwiched between two conducting media we find that, due to interference between waves reflected at the two interfaces, only certain types of waves can be propagated within the dielectric. To be more precise, let the dielectric be bounded by a perfect conductor at $y = 0$ and $y = a$; the thickness of the conductor will have no relevance to the problem because there is no penetration of the fields into the regions $y < 0$ or $y > a$.

Consider a plane wave incident at an angle $\theta$ to the plane $y = a$ (see Fig. 8). If it is polarized perpendicularly to the plane of incidence its electric vector will be

$$E = E_0 i e^{i(\omega t - ky \cos \theta - kz \sin \theta)}, \tag{3.55}$$

where $i$ is a unit vector in the $x$-direction. The electric vector of the reflected wave must be of the form

$$E = E_1 i e^{i(\omega t + ky \cos \theta - kz \sin \theta)}, \tag{3.56}$$

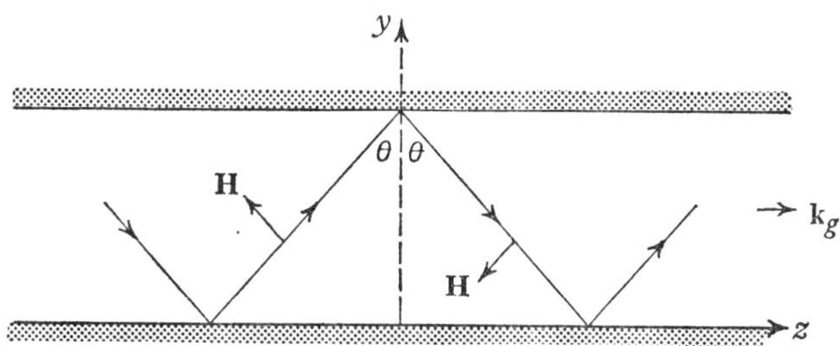


Fig. 8 A T.E. wave propagated between two conducting plates

To satisfy the boundary condition $E = 0$ at the conducting boundaries $y = 0$, $y = a$ for all $t$, $x$ and $z$, the incident and reflected waves must interfere destructively at both planes.

For $E = 0$ at $y = 0$

$$E_1 = -E_0. \tag{3.57}$$

For $E = 0$ at $y = a$

$$ka \cos \theta - n\pi, \tag{3.58}$$

where $n$ is an integer. Therefore, only waves incident in directions $\theta = \cos^{-1}(n\pi/ka)$ can be propagated between the planes. The electric vector of the resultant wave is

$$E = 2iE_0 \sin(k_c y) e^{i(\omega t - k_g z)}, \tag{3.59}$$

where $k_g = k \sin \theta$, $k_c = k \cos \theta$ and $k = \omega\sqrt{\varepsilon_1 \mu_1}$; they are related by

$$k^2 = k_c^2 + k_g^2. \tag{3.60}$$

This wave is propagated in the $z$-direction with a phase velocity $v_0 = \dfrac{\omega}{k_g} = \dfrac{vk}{k_g}$ where $v$ is the velocity of light in the dielectric. For free space, $v = c$ and so the phase velocity is greater than the velocity of light because $k > k_g$. However, the group velocity $\dfrac{d\omega}{dk_g}$, which can be calculated from (3.60). is given by

$$\frac{d\omega}{dk_g} = \frac{k_g v}{k},\tag{3.61}$$

which is always less than $c$.

The wave modes that exist can be classified by the values of $n$ in (3.58). If $\lambda_c$ and $\lambda_g$ are the wavelengths corresponding to wave vectors $k_c$ and $k_g$, from (3.60),

$$\lambda_c = \frac{2a}{n}, \quad \lambda_g = \frac{2\pi}{\omega}\left\{\varepsilon_1\mu_1 - \left(\frac{n\pi}{\omega a}\right)^2\right\}^{-\frac{1}{2}}.\tag{3.62}$$

If the $n$th mode is to be propagated without attenuation $\lambda_g$ must be real so

$$\omega > \frac{\pi n}{a\sqrt{\varepsilon_1\mu_1}}.\tag{3.63}$$

Therefore, for each mode there is a cut-off frequency

$$\omega_c = n\pi/a\sqrt{\varepsilon_1\mu_1},$$

since for frequencies less than this the mode will be attenuated. The smallest value of $n$ is unity so no wave can be propagated without attenuation which has a frequency less than $\pi/a\sqrt{\varepsilon_1\mu_1}$.

From Fig. 8 it is clear that through the electric vector is always perpendicular to the direction of propagation, the magnetic vector $H$, though perpendicular to $E$, has a component in the direction of propagation so this wave is a transverse electric wave. A completely independent type of wave with similar characteristics is obtained by considering the reflection of a wave polarized in the plane of incidence. This will be a transverse magnetic wave because $H$ will have no component in the $z$-direction.

### 3.3 Wave guides

The features of wave (3.57) are typical of waves propagated in a dielectric cylinder bounded by perfectly conducting walls. Such a system is known as a wave guide and is of considerable practical importance in the transmission of microwaves.

We shall consider the problem of wave propagation in wave guides in a fairly general way. If a wave of frequency $\omega$ is to be propagated in the $z$-direction along the axis of the cylinder, its electric and magnetic vectors must be of the form

$$E(x, y, z) = E_0(x, y)e^{i(\omega t - k_g z)}, \qquad (3.64)$$

$$H(x, y, z) = H_0(x, y)e^{i(\omega t - k_g z)}, \qquad (3.65)$$

where $k_g$ is known as the characteristic wave vector of the guide. $E$ and $H$ must satisfy the wave equation (2.11) and this condition leads to equations for $E_0$ and $H_0$,

$$\left(\frac{\partial^2}{\partial x^2} + \frac{\partial^2}{\partial y^2} + k_c^2\right)\begin{Bmatrix} E_0 \\ H_0 \end{Bmatrix} = 0, \qquad (3.66)$$

where $k = \omega\sqrt{\varepsilon\mu}$, for a dielectric $\varepsilon$, $\mu$, and $k_c^2 = k^2 - k_g^2$.

Further conditions on $E_0$ and $H_0$ are found by substituting the solutions of (3.64) into Maxwell's equations. Finally, the boundary conditions on the fields must be satisfied before the problem is completely solved.

In order to keep the rather tedious manipulation of equations to a minimum, we use Maxwell's equations to relate the transverse components of $E_0$, $H_0$ to the longitudinal components $E_{0z}$, $H_{0z}$.

If $k \neq k_g$

$$E_{0x, y} = \frac{i\omega\mu}{k^2 - k_g^2}\left\{k_0 \wedge \nabla H_{0z} - \frac{k_g}{\omega\mu}\nabla E_{0z}\right\}_{x, y}, \qquad (3.67)$$

$$H_{0x, y} = \frac{-i\omega\varepsilon}{k^2 - k_g^2}\left\{k_0 \wedge \nabla E_{0z} + \frac{k_g}{\omega\varepsilon}\nabla H_{0z}\right\}_{x, y}, \qquad (3.68)$$

where $k_0$ is a unit vector in the $z$-direction.

The procedure now reduces to solving (3.66) for the longitudinal component $H_{0z}$ for a T.E. wave or $E_{0z}$ for a T.M. wave, then satisfying the boundary conditions for these on the cylinder, and lastly determining the transverse components from (3.67) and (3.68). The appropriate boundary conditions on $E_{0z}$, $H_{0z}$ at the cylinder are

$$E_{0z} = 0 \quad (a) \qquad \frac{\partial H_{0z}}{\partial n} = 0 \quad (b), \qquad (3.69)$$

where $n$ is the direction normal to a point on the surface of the cylinder.

The condition on $E_{0z}$ follows directly from (1.47) because it is always tangential at the surface of the cylinder. The condition on

$H_{0z}$ can be shown by taking the component of (3.67) along the tangential direction to the cross-section of the cylinder

$$E_{0t} = \frac{i\omega\mu}{k^2 - k_g^2} \left\{ k_0 \wedge \nabla H_{0z} - \frac{k_g}{\omega\mu} \nabla E_{0z} \right\}_t. \qquad (3.70)$$

On the cylinder $E_{0t} = 0$ from (1.47), and $(\nabla E_{0z})_t = 0$ because $E_{0z}$ is zero at all points on $C$. From (3.68) it is clear that $H_{0z}$ has no component in the direction of the normal to the surface of the cylinder and hence (3.69$b$) follows.

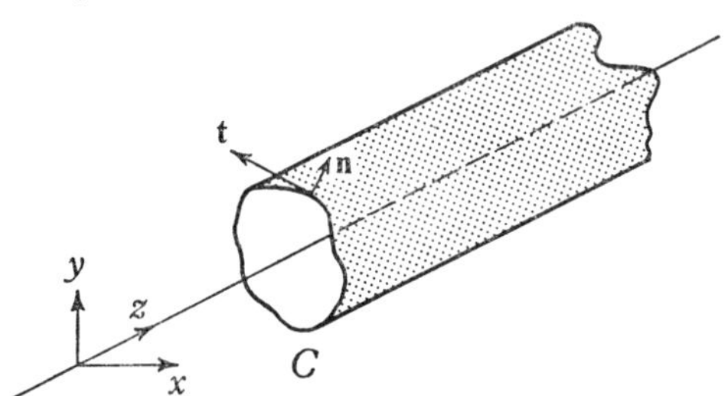


Fig. 9 The section of cylindrical wave guide generated by a closed curve $C$

In some wave guides a completely transverse wave, known as the T.E.M. or transverse electric and magnetic wave, can be propagated. On substituting (3.64) and (3.65) into Maxwell's equations and using the conditions $E_{0z} = 0$, $H_{0z} = 0$, it follows directly that $k = k_g$, in other words, the wave must be propagated with the same velocity as waves in the unbounded dielectric. This type of wave is important because it can be propagated without attenuation at all frequencies. The conditions on the boundary of a single component wave guide are too restrictive for such a wave to be propagated. A wave guide consisting of at least two components held at different potentials is needed. We can see this by considering the equation for $E_0(x, y)$. For $k = k_g$ this equation reduces to the equation for a static two dimensional field which implies that $E_0(x, y)$ can be calculated from a scalar potential function $\phi(x, y)$ which satisfies Laplace's equation $\nabla^2 \phi = 0$, and is related to $E_0$ by $E_0 = -\nabla\phi$. For a single component wave guide the only solution that satisfies Laplace's equation and the boundary condition $\phi = $ constant on the conductor, is $\phi = $ constant over the cross-section of the guide. This is the same argument that is used to prove the absence of an electric field inside a hollow conductor. A two component wave guide, such as two concentric circular cylinders, has a cross-section with two boundaries and if the potentials on the two components differ there may be a non-trivial solution for the potential that gives rise to electric and magnetic fields.

For T.E. and T.M. waves the boundary conditions (3.69) can be satisfied for certain discrete values of $k_c$ only. We shall show this for the particular case of a rectangular guide and indicate why it is true in general.

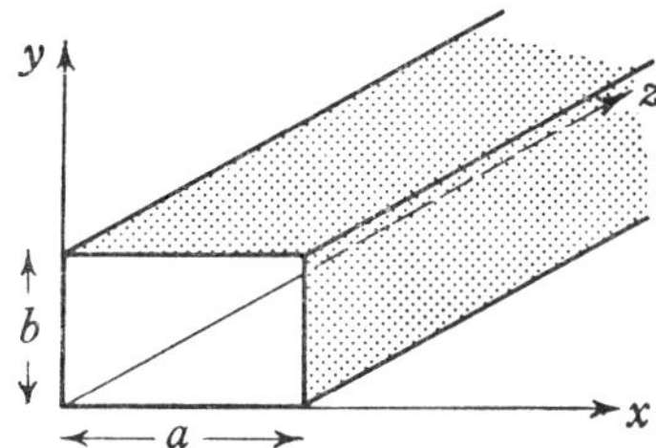

Fig. 10 A rectangular wave guide

Consider a wave guide formed by the conducting planes $x = 0$, $x = a$, $y = 0$ and $y = b$. For a T.M. wave we need to find a solution of (3.66) for $E_{0z}$ which will satisfy the boundary conditions. If we look for a product solution in Cartesian coordinates $E_{0z} = X(x)Y(y)$, (3.66) can be separated into two equations

$$\frac{d^2 X}{dx^2} + \lambda_1^2 X = 0, \quad \frac{d^2 Y}{dy^2} - \lambda_2^2 Y = 0, \tag{3.71}$$

where $\lambda_1^2$, $\lambda_2$ are separation constants which are related by the condition $\lambda_1^2 + \lambda_2^2 = k^2 - k_g^2$. The solutions of (3.69) result in an expression for $E_{0z}$ of the form:

$$E_{0z} = A \sin(\lambda_1 x + \delta_1) \sin(\lambda_2 y + \delta_2), \tag{3.72}$$

where $A$, $\delta_1$ and $\delta_2$ are arbitrary constants. The boundary conditions that $E_{0z}$ should vanish for all $t$ on the planes $x = 0$, $y = 0$, $x = a$, and $y = b$ means that $\delta_1 = 0$, $\delta_2 = 0$, and $\lambda_1$ and $\lambda_2$ are given by

$$\lambda_1 = \frac{n\pi}{a}, \quad \lambda_2 = \frac{m\pi}{b}, \tag{3.73}$$

where $n$ and $m$ are integers. Hence,

$$E_{0z} = A_{nm} \sin\left(\frac{n\pi x}{a}\right) \sin\left(\frac{n\pi y}{b}\right). \tag{3.74}$$

The transverse components can be calculated from (3.66) and (3.67).

For the $n, m$ transverse magnetic mode

$$k^2 - k_g^2 = \left(\frac{n\pi}{a}\right)^2 + \left(\frac{m\pi}{b}\right)^2. \tag{3.75}$$

The condition that $k_g$ should be real gives the condition on the frequency for the $n, m$ mode to be propagated without attenuation, which is

$$\omega > v \sqrt{\left(\frac{n\pi}{a}\right)^2 + \left(\frac{m\pi}{b}\right)^2}. \tag{3.76}$$

The coefficients $A_{nm}$ in the general solution for the electric field of a propagated wave,

$$E_z = \sum_{n,m} A_{nm} \sin\left(\frac{n\pi x}{a}\right) \sin\left(\frac{m\pi y}{b}\right) e^{i(\omega t - k_g z)}, \tag{3.77}$$

depend upon the way the wave is initially excited.

If the wave guide is closed by conducting planes at $z = 0$ and $z = d$ it becomes a resonance cavity. The waves are reflected by the end planes and standing waves that satisfy the boundary condition $E_x = 0$, $E_y = 0$ on the planes $z = 0$, $z = d$ are set up. This condition can be satisfied only for discrete values of $k_g$ such that

$$k_g = \frac{l\pi}{d}, \tag{3.78}$$

where $l$ is an integer. The form of $E_z$ that satisfies all these boundary conditions is

$$E_2 = \sum_{n,m,e} A nme \sin\left(\frac{n\pi x}{a}\right) \sin\left(\frac{m\pi y}{b}\right) \cos\left(\frac{l\pi z}{z}\right) e^{i\omega t}. \tag{3.79}$$

These equations are typical of the propagation in any cylindrical wave guide in that equation (3.79) are the homogeneous Sturm-Liouville type which must have discrete eigenvalues when boundary conditions of the form (3.69) are applied. The field vectors are expressed as a linear combination of modes corresponding to the allowed frequencies and they are determined completely only when the initial conditions are specified.

We have discussed here the propagation in an idealized wave guide, infinite in length and with perfectly conducting walls. In any real wave-guide system, account must be taken of attenuation due to energy dissipation in walls of finite conductivity, and reflections from the ends. Other problems which occur in real wave-guide systems are the coupling of wave guides, the injection and extraction of energy from guides, and problems of the design of guides required to propagate particular frequency bands.

## PROBLEMS 3

1. Find the form of the T.E. and T.M. modes that can be propagated along a cylindrical wave guide which has a circular cross-section of radius $a$ and perfectly conducting walls.

2. An harmonic wave of frequency $\omega$ is propagated without attenuation along a cylindrical wave guide which has perfectly conducting walls. Show that the time average of the Poynting vector along the axis is

$$\frac{k_g \omega \varepsilon}{2(k^2 - k_g^2)} \left\{ \left| \frac{\partial E_{0z}}{\partial x} \right|^2 + \left| \frac{\partial E_{0z}}{\partial y} \right|^2 \right\}$$

for a T.M. mode, and

$$\frac{k_g \omega \mu}{2(k^2 - k_g^2)} \left\{ \left| \frac{\partial H_{0z}}{\partial x} \right|^2 + \left| \frac{\partial H_{0z}}{\partial y} \right|^2 \right\}$$

for a T.E. mode. The notation is the same as that used in 3.3.

3. A wave guide, which has a square cross-section and perfectly conducting walls, has its axis in the $z$-direction and is filled with a dielectric $(\varepsilon_1, \mu_1)$ in the section $z<0$ and a dielectric $(\varepsilon_2, \mu_2)$ for $z>0$. If a T.E. mode $n = 1$ $m = 1$ is propagated from the negative $z$-direction, calculate the reflection and transmission coefficients at $z = 0$.

CHAPTER 4

# The Inhomogeneous Wave Equation

## 4.1 Electromagnetic potentials

Charges and currents are the source of all electromagnetic fields and, in certain circumstances, they will generate electromagnetic waves. To describe the emission of an electromagnetic wave we must retain the terms $\rho$ and $J$ in Maxwell's equations and use them in the form (1.1) to (1.4). These equations are more difficult to handle than the homogeneous equations (2.1) to (2.4) so it is of considerable advantage to reduce the number of equations by introducing some form of potential function. The functions that are most commonly used are the vector potential $A$ and the scalar potential $\phi$. The vector $A$ is chosen such that

$$B = \operatorname{curl} A. \tag{4.1}$$

This form for $B$ satisfies equation (1.2) identically. Any solution of (1.2) which has continuous first order partial derivatives can be expressed as the curl of some vector, so (4.1) is perfectly general. If this form is substituted into (1.4), the equation can be rewritten as

$$\operatorname{curl}\left(E + \frac{\partial A}{\partial t}\right) = 0. \tag{4.2}$$

Any vector which has a zero curl and has continuous first order partial derivatives can be written as the gradient of some scalar function. We can, therefore, introduce $\phi$ such that

$$E = -\nabla\phi - \frac{\partial A}{\partial t}. \tag{4.3}$$

The equations (4.1) and (4.3) do not define $A$ and $\phi$ uniquely. New potentials $A'$, $\phi'$ can be constructed, which give the same fields $E$ and $B$, by

$$A' = A - \nabla\psi, \quad \phi' = \phi + \frac{\partial\psi}{\partial t}, \tag{4.4}$$

where $\psi$ is an arbitrary scalar function. The transformation

$$(A, \phi) \rightarrow (A', \phi')$$

52

is known as a *gauge transformation* and, provided that no further conditions are placed on the potentials, all expressions for the fields should be invariant under such a transformation.

Equations for $A$ and $\phi$ can be found by substituting $B$ and $E$ from (4.1) and (4.3) into (1.1) and (1.3). If $D = \varepsilon E$ and $B = \mu H$, these equations can be written in the form

$$\nabla^2 A - \mu\varepsilon \frac{\partial^2 A}{\partial t^2} = -\mu J - \operatorname{grad}\left(\operatorname{div} A + \mu\varepsilon \frac{\partial \phi}{\partial t}\right), \tag{4.5}$$

and

$$\nabla^2 \phi - \mu\varepsilon \frac{\partial^2 \phi}{\partial t^2} = -\rho/\varepsilon - \frac{\partial}{\partial t}\left(\operatorname{div} A + \mu\varepsilon \frac{\partial \phi}{\partial t}\right). \tag{4.6}$$

Only the curl of the vector $A$ has been defined so far and we can specify the divergence independently (see Helmholtz Theorem, Appendix B). We can define it by the condition

$$\operatorname{div} A + \mu\varepsilon \frac{\partial \phi}{\partial t} = 0, \tag{4.7}$$

so as to simplify (4.5) and (4.6).

This further restriction on $A$ is known as the Lorentz gauge condition. The potentials $A$ and $\phi$ are still not uniquely determined because $A'$, $\phi'$, defined by (4.4), satisfy (4.7) if $\psi$ is *any* solution of the homogeneous scalar wave equation.

In the Lorentz gauge the equations for $A$ and $\phi$ are independent:

$$\nabla^2 A - \mu\varepsilon \frac{\partial^2 A}{\partial t^2} = -\mu J, \tag{4.8}$$

and

$$\nabla^2 \phi - \mu\varepsilon \frac{\partial^2 \phi}{\partial t^2} = -\rho/\varepsilon. \tag{4.9}$$

Thus both potentials $A$ and $\phi$ satisfy inhomogeneous wave equations.

An alternative potential to $A$ and $\phi$ is the Hertz potential $Z(r, t)$. This potential is chosen to satisfy the equations

$$A = \mu\varepsilon \frac{\partial Z}{\partial t}, \quad \phi = -\operatorname{div} Z, \tag{4.10}$$

so that the Lorentz gauge condition (4.7) is satisfied identically. Given $A$ and $\phi$, it can be proved that a $Z$ satisfying (4.10) can always be found (see problem 1). The solution for $Z$ is not unique and possible solutions can differ by curl $g$, where $g$ is a vector constant in time.

By the same argument a vector $p$ can be introduced by the equations

$$J = \frac{\partial p}{\partial t}, \quad \rho = -\text{div } p, \tag{4.11}$$

which are compatible with the charge continuity equation (1.18). The vector $p$ defined in this way is not unique to the extent of the addition of the curl of an arbitrary time independent vector.

It follows, on substituting (4.10) and (4.11) into (4.8) and (4.9), that $Z$ must satisfy the equation

$$\nabla^2 Z - \mu\varepsilon \frac{\partial^2 Z}{\partial t^2} = -\frac{p}{\varepsilon} + \text{curl} f, \tag{4.12}$$

where $f$ is an arbitrary vector independent of time. We can effectively put curl $f = 0$, by making use of the fact that $Z$ is arbitrary to the extent of the addition of a curl $g$, and eliminate $f$ by choosing $g$ such that $\nabla g = f$.

From (4.10), (4.1) and (4.3) we can express the fields $E$ and $H$ in terms of $Z$ by

$$B = \mu\varepsilon \text{ curl } \frac{\partial Z}{\partial t}, \tag{4.13}$$

and

$$E = \text{grad div } Z - \mu\varepsilon \frac{\partial^2 Z}{\partial t^2} \tag{4.14}$$

It is possible to interpret $p$ as the electric dipole moment per unit volume associated with the charge and current sources $\rho$ and $J$. From the first equation of (4.11) we have $p = \int J \partial t$. If we consider $N$ charges per unit volume $q_i$ moving with velocities $v_i$, the associated current density $J$ is $\sum_{i=1}^{N} q_i v_i$. If this is integrated with respect to $t$ we find $p = \sum_{i=1}^{N} q_i r_i$, which, apart from a constant with respect to time, is the electric dipole moment per unit volume.

### 4.2 Kirchhoff's solution

There are various ways of generating a solution of a partial differential equation. Most problems require a solution that will satisfy specific initial and boundary conditions. These can generally be satisfied by a certain type of solution only, which means that the boundary conditions often dictate the method of solution of the equation. For the wave equation Kirchhoff (1824–87) derived an important general solution of (4.9) for $\phi$ at points within a volume $V$ bounded by a surface $S$, given values of the potential $\phi$ and its normal derivatives on $S$. The method of integration is based on the use of Green's Theorem,

$$\int_V (\psi_1 \nabla^2 \psi_2 - \psi_2 \nabla^2 \psi_1) d\tau = \int_S \left( \psi_1 \frac{\partial \psi_2}{\partial n} - \psi_2 \frac{\partial \psi_1}{\partial n} \right) dS, \quad (4.15)$$

where $\psi_1$ and $\psi_2$ are arbitrary scalar functions which must be continuous and have continuous first and second order partial derivatives within $V$ and on $S$. The vector $n$ is a unit vector along the outward normal and $\dfrac{\partial \psi_1}{\partial n}$ denotes the derivatives of $\psi_1$ in this direction. The reader may be familiar with this method of solution as applied to Laplace's and Poisson's equations.

It will be convenient to work with the equation for a single frequency component only. This we can do by taking the Fourier transform of (4.9), which leads to the equation

$$\nabla^2 \phi_\omega + \frac{\omega^2}{v^2} \phi_\omega = - \frac{\rho_\omega}{\varepsilon}, \quad (4.16)$$

where $\rho_\omega$ is the Fourier transform of the charge density given by

$$\rho_\omega(r) = \frac{1}{\sqrt{2\pi}} \int_{-\infty}^{\infty} \rho(r, t)e^{-i\omega t}dt, \quad (4.17)$$

and $\phi_\omega(r)$ is the Fourier transform of $\phi(r, t)$.

Given the form of $\phi_\omega$ and $\dfrac{\partial \phi_\omega}{\partial n}$ on $S$, the aim is to find the solution $\phi_\omega$ of (4.16) at all points within $V$. As a preliminary we find the potential $\psi_\omega$ due to a point source at any point P within $V$. The potential $\psi_\omega$ must satisfy the homogeneous equation

$$\nabla^2 \psi_\omega + \frac{\omega^2}{v^2} \psi_\omega = 0 \quad (4.18)$$

at all points in $V$ except $P$. The form of $\psi_\omega$ for a point source must be a function of $R$ only, where $R$ is the radial distance from the point $P$.

In terms of $R$, (4.16) simplifies to

$$\frac{d^2\psi_\omega}{dR^2} + \frac{2}{R}\frac{d\psi_\omega}{dR} + \frac{\omega^2}{v^2}\psi_\omega = 0. \qquad (4.19)$$

For $\omega = 0$, (4.16) reduces to Laplace's equation and we know that the appropriate solution is $1/4\pi\varepsilon R$, which is the potential due to a unit point charge at the origin. For $\omega \neq 0$, the equation can be written in terms of $R\psi_\omega$ as

$$\frac{d^2}{dR^2}(R\psi_\omega) + \frac{\omega^2}{v^2}(R\psi_\omega) = 0. \qquad (4.20)$$

There are two independent solutions for $\psi_\omega$, one is an outgoing wave $e^{-i\frac{\omega R}{v}}/4\pi\varepsilon R$ when combined with the time factor $e^{i\omega t}$, the other is an incoming wave $e^{i\frac{\omega R}{v}}/4\pi\varepsilon R$.

To integrate (4.16) we shall use Green's Theorem and choose $\psi_1$ as $\phi_\omega$ and $\psi_2$ as the unit source term $e^{-i\frac{\omega}{v}|\mathbf{r}-\mathbf{r}'|}/4\pi\varepsilon|\mathbf{r}-\mathbf{r}'|$, where $\mathbf{r}$ is the position vector of $\mathbf{P}$. The unit source term has the effect of isolating the potential $\phi_\omega$ at $\mathbf{P}$ and relating it to $\phi_\omega$ and its derivative on $S$. The details of the argument are rather involved, because $\mathbf{P}$ must be excluded as $\psi_2$ is not continuous at this point. A small spherical surface $S'$ centred on $\mathbf{P}$ with a radius $\delta$ is used to remove $\mathbf{P}$ and Green's Theorem is applied to the volume bounded by this surface $S'$ and the outer surface $S$. The contribution to the surface integral of (4.14) from $S'$ is

$$\frac{1}{4\pi\varepsilon}\int_0^{2\pi}\int_0^{\pi} e^{-\frac{i\omega\delta}{v}}\left(\phi_\omega + \frac{i\omega\delta\phi_\omega}{v} + \delta\frac{\partial\phi_\omega}{\partial n}\right)\sin\theta d\theta d\psi. \qquad (4.21)$$

In the limit $\delta \to 0$ the potential at P is isolated and the integral is equal to $\phi_\omega(r)/\varepsilon$ provided $\phi_\omega$ and $\dfrac{\partial\phi_\omega}{\partial n}$ are bounded on $S'$.

The other contribution to the surface integral in (4.15) is from $S$ and is equal to

$$\frac{1}{4\pi\varepsilon}\int_S\left\{\phi_\omega\frac{\partial}{\partial n}\left(\frac{e^{-i\frac{\omega}{v}|\mathbf{r}-\mathbf{r}'|}}{|\mathbf{r}-\mathbf{r}'|}\right) - \frac{e^{-i\frac{\omega}{v}|\mathbf{r}-\mathbf{r}'|}}{|\mathbf{r}-'\mathbf{r}|}\frac{\partial\phi_\omega}{\partial n}\right\}ds. \qquad (4.22)$$

The volume integral on the right hand side of (4.15) simplifies to

$$\frac{1}{4\pi\varepsilon}\int_V\frac{\rho_\omega e^{-i\frac{\omega}{v}|\mathbf{r}-\mathbf{r}'|}}{\varepsilon|\mathbf{r}-\mathbf{r}'|}d\tau' \qquad (4.23)$$

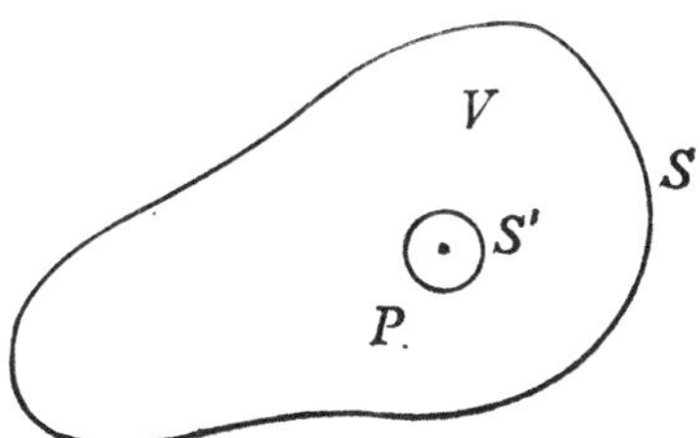

Fig. 11 A closed surface $S$ bounding a
volume $V$. $P$ is a point in $V$. $S'$ is a small
sphere centred on $P$, which has a radius
$\delta$, and lies completely within $S$

because $\phi_\omega$ satisfies (4.16) and $\psi_\omega$ satisfies (4.18). The contributions
(4.21), (4.22) and (4.23) to Green's Theorem can be rearranged to
give an expression for $\phi_\omega(r)$

$$\phi_\omega(r) = \frac{1}{4\pi\varepsilon} \int_V \frac{\rho_\omega e^{-i\frac{\omega}{v}|r-r'|}}{|r-r'|}\, d\tau' - \frac{1}{4\pi} \int_S \left\{ \phi_\omega \frac{\partial}{\partial n}\left(\frac{1}{|r-r'|}\right) \right.$$

$$\left. - \frac{i\omega\phi_\omega}{v|r-r'|}\frac{\partial|r-r'|}{\partial n} - \frac{1}{|r-r'|}\frac{\partial\phi_\omega}{\partial n} \right\} e^{-i\frac{\omega}{v}|r-r'|}\, dS \quad (4.24)$$

and this is a solution of (4.16) which was obtained by Helmholtz
(1821-94).

Kirchhoff's solution of (4.9) can be obtained by making a Fourier
inversion of (4.24). The exponential factor $e^{-\frac{i\omega}{v}|r-r'|}$ has the effect
of giving the functions a dependence upon $t - |r-r'|/v$. For example,
the Fourier inverse of the first term on the right-hand side of (4.24) is

$$\frac{1}{\sqrt{2\pi}} \int_{-\infty}^{\infty} \frac{1}{4\pi\varepsilon} \int_V \frac{\rho_\omega e^{i\omega\left(t - \frac{|r-r'|}{v}\right)}}{|r-r'|}\, d\tau'\, d\omega. \quad (4.25)$$

If the order of integration is reversed and the $\omega$-integration is
carried out, (4.25) becomes

$$\frac{1}{4\pi\varepsilon} \int_V \frac{\rho\left(r', t - \frac{|r-r'|}{v}\right)}{|r-r'|}\, d\tau'. \quad (4.26)$$

The Fourier inverse of the complete equation (4.24) is

$$\phi(r, t) = \frac{1}{4\pi\varepsilon} \int_V \frac{[\rho]}{|r-r'|}\, d\tau' - \frac{1}{4\pi} \int_S \left\{ [\phi]\frac{\partial}{\partial n}\left(\frac{1}{|r-r'|}\right) \right.$$

$$\left. - \frac{1}{v|r-r'|}\left[\frac{\partial\phi}{\partial t}\right]\frac{\partial|r-r'|}{\partial n} - \frac{1}{|r-r'|}\left[\frac{\partial\phi}{\partial n}\right] \right\} dS \quad (4.27)$$

which is Kirchhoff's solution of (4.9). The first term on the right-hand side is the particular integral and a solution of the inhomogeneous equation. The surface integral over $S$ is a complementary function and a solution of the homogeneous equation.

The square brackets are used to indicate that the function contained within them must be evaluated at time $t - \dfrac{|\mathbf{r}-\mathbf{r'}|}{v}$. The potential depends upon the charge density, $\phi$, $\dfrac{\partial \phi}{\partial t}$ and $\dfrac{\partial \phi}{\partial n}$ at an earlier time because information takes a finite time $\dfrac{|\mathbf{r}-\mathbf{r'}|}{v}$ to travel from a source at $\mathbf{r'}$ to the point $\mathbf{r}$. For this reason (4.27) is known as the *retarded potential*. An independent solution can be obtained by the same method of integration but by using the incoming wave $e^{-i\frac{\omega}{v}|\mathbf{r}-\mathbf{r'}|}/4\pi\varepsilon\,|\mathbf{r}-\mathbf{r'}|$ for $\psi_2$. This solution is of the same form as (4.27), except that the terms are evaluated at a later time $t + \dfrac{|\mathbf{r}-\mathbf{r'}|}{v}$, and it is therefore known as the *advanced potential*. It is usually rejected on the physical grounds that it conflicts with the Principle of Causality as it implies that a field can be observed even before it has been generated.

A similar solution to (4.27) can be derived for the potential *outside* a closed surface $S$. Consider a finite distribution of charge completely surrounded by a spherical surface $S''$. If $S''$ is made large enough to enclose the surface $S$ and the sphere $S'$ which surrounds the point of observation $P$, then Green's Theorem can be applied to the volume $V$ bounded by these three surfaces. The integrals over $S'$ and $S$ are the same as in (4.21) and (4.22) respectively. If $S''$ is centred at P and has a radius $R$, the surface integral over $S''$ is

$$\frac{1}{4\pi\varepsilon}\int_0^{2\pi}\int^{\pi} e^{-\frac{i\omega R}{v}}\left\{\phi_\omega + R\frac{\partial\phi_\omega}{\partial R} + \frac{i\omega}{v}R\phi_\omega\right\}\sin\theta\,d\theta\,d\psi. \qquad (4.28)$$

If $R\phi_\omega(R)$ is bounded, and if

$$R\left(\frac{\partial\phi_\omega}{\partial R} + \frac{i\omega}{v}\phi_\omega\right) \to 0 \qquad (4.29)$$

as $R \to \infty$ uniformly with respect to direction, this contribution to the surface integral goes to zero in the limit $R \to \infty$. If these conditions hold, the retarded solution for the potential at $P$ is

$$\phi_\rho(r, t) = \frac{1}{4\pi\varepsilon} \int_V \frac{[\rho]}{|r - r'|} \, d\tau + \frac{1}{4\pi} \int_S \left\{ [\phi] \frac{\partial}{\partial n} \left( \frac{1}{|r - r'|} \right) \right.$$

$$\left. - \frac{1}{v|r - r'|} \left[ \frac{\partial\phi}{\partial t} \right] \frac{\partial |r - r'|}{\partial n} - \frac{1}{|r - r'|} \left[ \frac{\partial\phi}{\partial n} \right] \right\} dS. \quad (4.30)$$

The condition (4.29) means essentially that the solution must behave as an outgoing wave in the limit $R \to \infty$ It is often known as 'Sommerfeld's radiation condition' and in some problems must be imposed as a boundary condition at infinity to obtain a unique solution.

If we reconsider solution (4.27) and choose $S$ to be a large sphere of radius $R$, in the limit $R \to \infty$ we find

$$\phi(r, t) = \frac{1}{4\pi\varepsilon} \int_V \frac{[\rho(r', t)]}{|r - r'|} \, d\tau', \quad (4.31)$$

provided $\phi$ satisfies the radiation condition (4.29) so that the surface integral makes no contribution to $\phi$. This form for the potential is similar to that for a static charge distribution, except that the factor $1/4\pi\varepsilon |r - r'|$ is weighted by a *retarded charge distribution*. One might be tempted to assume that the potential due to a point charge $q$ is $q/4\pi\varepsilon |r - r'|$, with $r'$ referring to the retarded position of the charge. However we shall see in the next section that this is not correct because the *apparent* total charge $\int_V [\rho(r', t)] d\tau'$ of the retarded distribution is not in general equal to the total real charge.

Similar considerations apply in solving the vector wave equation (4.8) for the potential $A$. If $A$ satisfies a radiation condition similar to (4.29) then (4.8) has the unique solution,

$$A(r, t) = \frac{\mu}{4\pi} \int_V \frac{[j(r', t)]}{|r - r'|} \, d\tau', \quad (4.32)$$

where the integration extends over all space.

### 4.3 The potentials due to a moving point charge

Our aim in this section is to evaluate the retarded potentials (4.29) and (4.30) for a point charge whose motion is governed by the equation $r = r_0(t)$. The concept of a point charge is very useful, because from a knowledge of the fields associated with a point charge, those associated with a more complicated charge distribution can often be found. A linear combination of fields is taken which corresponds to point charges which build up to the particular charge distribution under consideration. There are, however, difficulties in finding a suitable mathematical description of a point charge. It seems reasonable to assume that the potential at large distances from a small charge which can be surrounded by a small sphere $\Sigma$ of radius $\partial$ should depend only on the total charge $q$, and not on the way the charge is distributed within $\Sigma$, if $\partial$ is small enough. If this is true, then any charge distribution within $\Sigma$ will give rise to the same form for the potentials $A$ and $\phi$ in the limit $\partial \to 0$. The most convenient distribution to take is one in which the charge is evenly distributed throughout $\Sigma$, so that $\rho(r, t) = qf(r)$, where

$$f(r) = \frac{3}{4\pi\partial^3}, \qquad |r-r_0| \leq \partial,$$

$$= 0, \qquad |r-r_0| > \partial. \tag{4.33}$$

The current density associated with this distribution is $qu(t)f(r)$, where $u(t) = \dfrac{dr_0}{dt}$.

From (4.9) the scalar potential of this distribution is

$$\phi(r, t) = \frac{3q}{16\pi^2\varepsilon\partial^3} \int_V \frac{[f(r')]}{|r-r'|} \, d\tau'. \tag{4.34}$$

A light signal from the point $r$ takes a time $|r-r'|/v$ to reach the point of observation $P$, which has the position vector $r$. When this signal was emitted the centre of the charge must have been in the position $r_0\left(t - \dfrac{|r-r'|}{v}\right)$. It will be convenient to measure distances from this position and change the variable of integration from $r'$ to $s$, where

$$s = r' - r_0\left(t - \frac{|r-r'|}{v}\right). \tag{4.35}$$

The Jacobian of the transformation $\dfrac{\partial(s_1, s_2, s_3)}{\partial(x', y', z')}$ can be calculated

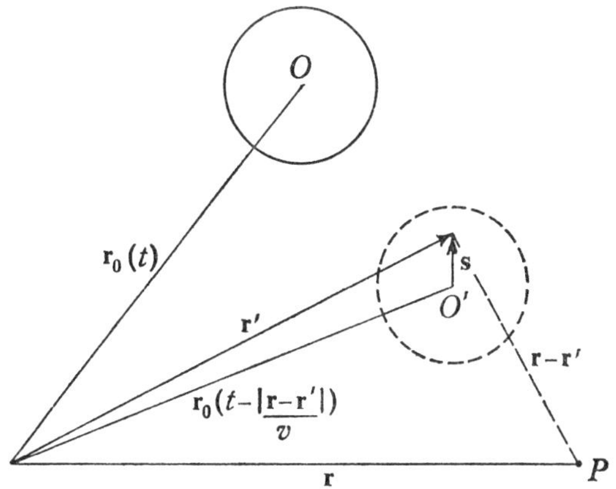


Fig. 12  $O$ is the position of the centre of the charge at time $t$. $O'$ is the position of the centre when a light signal, emitted from $\mathbf{r}'$, reaches the observation point $P$ at the moment $t$

from (4.35) by taking the appropriate derivatives. The $x$-component of (4.35) is

$$s_1 = x' - x_0\left(t - \frac{|\mathbf{r}-\mathbf{r}'|}{v}\right).$$

(4.36)

On differentiating this with respect to $x'$, we find

$$\frac{\partial s_1}{\partial x'} = 1 - \frac{\partial x_0}{\partial x'}.$$

(4.37)

Because $x_0\left(t - \dfrac{|\mathbf{r}-\mathbf{r}'|}{v}\right)$ has the same functional dependence upon

$t$ as upon $-\dfrac{|\mathbf{r}-\mathbf{r}'|}{v}$ we have the relation

$$\frac{\partial x_0}{\partial x'} = -\frac{1}{v}\frac{\partial x_0}{\partial t}\frac{\partial |\mathbf{r}-\mathbf{r}'|}{\partial x'}.$$

(4.38)

However,

$$\frac{\partial |\mathbf{r}-\mathbf{r}'|}{\partial x'} = \frac{x'-x}{|\mathbf{r}-\mathbf{r}'|},$$

(4.39)

and (4.38) (4.39) can be substituted into (4.37) to give

$$\frac{\partial s_1}{\partial x'} = 1 - \frac{u_x(x-x')}{v|\mathbf{r}-\mathbf{r}'|},$$

(4.40)

where

$$u_x = \frac{\partial x_0}{\partial t}.$$

Similarly, the other terms in the Jacobian can be evaluated, such as

$$\frac{\partial s_1}{\partial y'} = -\frac{u_x(y-y')}{v\,|\,r-r'\,|}, \qquad \frac{\partial s_1}{\partial z'} = -\frac{u_z(z-z')}{v\,|\,r-r'\,|}. \tag{4.41}$$

The final form for the Jacobian is

$$\begin{vmatrix} 1-\dfrac{u_x(x-x')}{v\,|\,r-r'\,|} & -\dfrac{u_x(y-y')}{v\,|\,r-r'\,|} & -\dfrac{u_x(z-z')}{v\,|\,r-r'\,|} \\[2ex] -\dfrac{u_y(x-x')}{v\,|\,r-r'\,|} & 1-\dfrac{u_y(y-y')}{v\,|\,r-r'\,|} & -\dfrac{u_y(z-z')}{v\,|\,r-r'\,|} \\[2ex] -\dfrac{u_z(x-x')}{v\,|\,r-r'\,|} & -\dfrac{u_z(y-y')}{v\,|\,r-r'\,|} & 1-\dfrac{u_z(z-z')}{v\,|\,r-r'\,|} \end{vmatrix} \tag{4.42}$$

The determinant can be expanded and the terms regrouped to express the result in the form

$$\frac{\partial(s_1,\,s_2,\,s_3)}{\partial(x',\,y',\,z')} = \left| \, 1-\frac{(r-r')\cdot u}{|\,r-r'\,|\,v} \, \right|. \tag{4.43}$$

The modulus is taken because the Jacobian must always be positive.

The transformed integral (4.34) for $\phi(r,\,t)$ becomes

$$\phi(r,\,t) = \frac{3q}{16\pi^2\varepsilon\delta^3} \int_\Sigma \frac{ds_1\,ds_2\,ds_3}{\left[\,|\,r-r_0-s\,| - \dfrac{(r-r_0-s)\cdot u}{v}\,\right]}, \tag{4.44}$$

where $r'$ has been expressed in terms of $r_0$ and $s$ from (4.35). The integral need not be evaluated explicitly to obtain the value in the limit $\delta\to0$. By the mean value theorem the integral is equal to the integrand evaluated at some point within the sphere $\Sigma$ multiplied by the volume $\frac{4\pi}{3}\delta^3$. In the limit $\delta\to0$ this is

$$\phi(r,\,t) = \frac{q}{4\pi\varepsilon} \frac{1}{\left[\,|\,r-r_0\,| - \dfrac{(r-r_0)\cdot u}{v}\,\right]}. \tag{4.45}$$

The square brackets must still be used because $r_0(t')$, and $u(t')$ have to be evaluated at the time $t'$ when the charge was at its retarded position. The time $t'$ can be calculated from

$$t' = t - \frac{|\,r - r_0(t')\,|}{v}.$$
(4.46)

Alternatively, the retarded position $r_e$ can be calculated from

$$r_e = r_0\left(t - \frac{|\,r - r_e\,|}{v}\right).$$
(4.47)

The corresponding expression for the vector potential of the charge $A$ is

$$A = \frac{\mu q}{4\pi}\left[\frac{u}{|\,r - r_0\,| - \dfrac{(r - r_0) \cdot u}{v}}\right].$$
(4.48)

The expressions (4.45) and (4.46) are known as the Liénard-Wiechert potentials.

An alternative technique of dealing with a point charge is to use the Dirac delta function $\delta(x - x')$ which has the property

$$\delta(x - x') = 0, \qquad x \neq x',$$
(4.49)

and

$$\int f(x)\delta(x - x')dx = f(x'),$$
(4.50)

where $f(x)$ is an arbitrary function and the integration is over an interval which includes the point $x$. There is no function in the normal sense which has the two properties (4.49) and (4.50), though a function of this type can be regarded as the limit of a sequence of functions and is known as a weak or generalized function. For the definition and properties of generalized functions we refer the reader to [3] in the bibliography. In terms of the $\delta$-function the distribution function of a point charge $q$ at the point $r_0$ is

$$\rho(r, t) = q\delta(x - x_0)\delta(y - y_0)\delta(z - z_0).$$
(4.51)

We shall use this form in some of the later sections, but we shall need only to refer to the properties (4.49) (4.50).

## 4.4 The fields due to a point charge moving with constant velocity

It is possible, in the case of a point charge moving with constant velocity, to solve the equation (4.46) and express the potentials in terms of the position of the charge at the time of observation $t$.

Let the charge move parallel to the $x$-axis with a constant speed $u(<v)$. Let $P$ be the point of observation with position vector $r$, let $A$ be the retarded position of the charge $r_e$, and $B$ the position $r_0$ at the time of observation. In the time $\Delta t$ taken by the charge to travel the distance from $A$ to $B$, a light signal has travelled from $A$ to $P$.

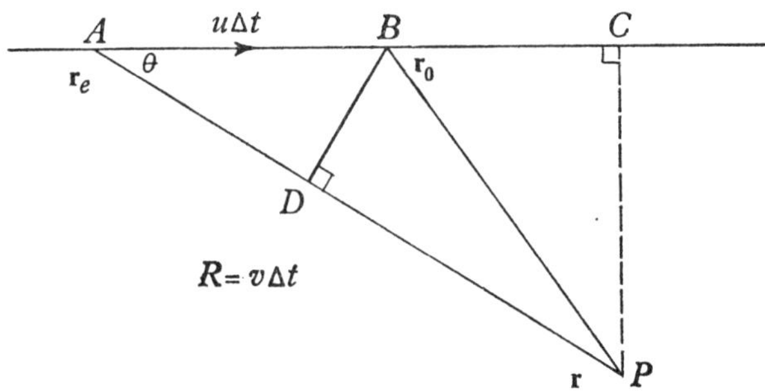


Fig. 13 A charge $q$ moves along $ABC$ with a constant velocity $u$. $A$ is the retarded position for the point $P$. $B$ is the actual position at the moment of observation

From Fig. 13

$$PD = R - u\Delta t \cos \theta, \tag{4.52}$$

where $R = |r - r_e|$.

However, $\Delta t = R/v$ and $\cos \theta = R \cdot u/uR$, and if these are substituted into (4.52) we find

$$PD = R - \frac{R \cdot u}{v} \tag{4.53}$$

which is the retarded denominator of the Liénard-Wiechert potentials (4.43) and (4.46). $PD$ can be calculated in terms of $r$ and $r_0$ by using the geometry of Fig. 12. From triangle $PBD$,

$$PD^2 = r_0^2 - u^2\Delta^2 t \sin^2 \theta, \tag{4.54}$$

but

$$\Delta t \sin \theta = \frac{PC}{v}, \tag{4.55}$$

and

$$PC^2 = (y - y_0)^2 + (z - z_0)^2. \tag{4.56}$$

On substituting (4.56) and (4.55) in (4.54) we obtain an equation for $PD$ in terms of $r_0$ the actual position of the charge, which is

$$PD^2 = (x - x_0)^2 + \left\{(y - y_0)^2 + (z - z_0)^2\right\}\left(1 - \frac{u^2}{v^2}\right). \tag{4.57}$$

Hence, the potentials $A$ and $\phi$ from (4.45) (4.48) are

$$\phi(r, t) = \frac{q}{4\pi\varepsilon} \frac{1}{\left\{(x-x_0)^2+\{(y-y_0)^2+(z-z_0)^2\}\left(1-\dfrac{u^2}{v^2}\right)\right\}^{\frac{1}{2}}} \qquad (4.58)$$

and

$$A(r, t) = \frac{\mu q}{4\pi} \frac{u}{\left\{(x-x_0)^2+\{(y-y_0)^2+(z-z_0)^2\}\left(1-\dfrac{u^2}{v^2}\right)\right\}^{\frac{1}{2}}}. \qquad (4.59)$$

It is now a relatively straightforward matter to calculate the fields from the potentials, using (4.1) and (4.3). If the charge passes through the origin at $t = 0$ then $x_0 = ut$, $y_0 = 0$, $z_0 = 0$, and the electric field is given by

$$E(r, t) = \frac{q}{4\pi\varepsilon} \frac{(r-r_0)\left(1-\dfrac{u^2}{v^2}\right)}{\left\{(x-ut)^2+(y^2+z^2)\left(1-\dfrac{u^2}{v^2}\right)\right\}^{\frac{3}{2}}}. \qquad (4.60)$$

It is directed along the line joining the actual position of the charge to the point of observation.

The magnetic induction field is perpendicular to $E$ and is given by

$$B(r, t) = \frac{\mu q}{4\pi} \frac{u \wedge (r-r_0)\left(1-\dfrac{u^2}{v^2}\right)}{\left\{(x-ut)^2+(y^2+z^2)\left(1-\dfrac{u^2}{v^2}\right)\right\}^{\frac{3}{2}}}. \qquad (4.61)$$

### 4.5 Fields due to an accelerated point charge

For a general equation of motion $r = r_0(t)$ it is not possible to express the potentials explicitly in terms of the actual position of the charge at the time of observation. We cannot solve the equation

$$r_e = r_0\left(t - \frac{|r-r_e|}{v}\right), \qquad (4.62)$$

for the retarded position $r_e$ in terms of $r$ and $t$, so the dependence of $r_e$ on $r$ and $t$ must be taken into account implicitly.

To calculate the fields $E$ and $B$ we shall need to know the derivatives of $R(= r-r_e)$ and $R$ with respect to the variables $x$, $y$, $z$ and $t$.

From (4.60), $R$ must satisfy the equation

$$R = r - r_0\left(t - \frac{R}{v}\right).\qquad(4.63)$$

If $R$ has cartesian components $(R_x, R_y, R_z)$ we can show by differentiating the first component of (4.63) with respect to $x$ that

$$\frac{\partial R_x}{\partial x} = 1 + \frac{u_x}{v}\frac{\partial R}{\partial x},\qquad(4.64)$$

where

$$u_x = \frac{dx_0}{dt}.$$

The derivative of $R$ is related to the derivatives of the components of $R$ by the equation

$$\frac{\partial R}{\partial x} = \frac{R_x\dfrac{\partial R_x}{\partial x} + R_y\dfrac{\partial R_y}{\partial y} + R_z\dfrac{\partial R_z}{\partial z}}{R}.\qquad(4.65)$$

By differentiating the second and third components of (4.63) with respect to $x$ we have

$$\frac{\partial R_y}{\partial x} = \frac{u_y}{v}\frac{\partial R}{\partial x} \quad\text{and}\quad \frac{\partial R_z}{\partial x} = \frac{u_z}{v}\frac{\partial R}{\partial x}.\qquad(4.66)$$

If we substitute (4.64) and (4.66) into (4.65) we obtain an equation for $\dfrac{\partial R}{\partial x}$. It can be rearranged to give

$$\frac{\partial R}{\partial x} = \frac{R_x}{\left(R - \dfrac{R\cdot u}{v}\right)}.\qquad(4.67)$$

This result can be substituted back into (4.64) and (4.66) to find the derivatives of the components of $R$ with respect to $x$. A similar argument can be used to calculate the derivatives with respect to $y$ and $z$.

Derivatives of $R$ and $R$ with respect to $t$ will also be needed. Differentiating (4.63) with respect to $t$ we find

$$\frac{\partial R}{\partial t} = -u\left(1 - \frac{1}{v}\frac{\partial R}{\partial t}\right).\qquad(4.68)$$

We also have the relation

$$\frac{\partial R}{\partial t} = \frac{\boldsymbol{R} \cdot \dfrac{\partial \boldsymbol{R}}{\partial t}}{R}. \tag{4.69}$$

From (4.68) and (4.69) we find the following expression for $\dfrac{\partial R}{\partial t}$:

$$\frac{\partial R}{\partial t} = \frac{-\boldsymbol{u} \cdot \boldsymbol{R}}{\left(R - \dfrac{\boldsymbol{R} \cdot \boldsymbol{u}}{v}\right)}. \tag{4.70}$$

Hence from (4.68) we find

$$\frac{\partial \boldsymbol{R}}{\partial t} = -\boldsymbol{u}\,\frac{R}{\left(R - \dfrac{\boldsymbol{R} \cdot \boldsymbol{u}}{v}\right)}. \tag{4.71}$$

We are now in a position to calculate the fields $E$ and $B$ from the Liénard-Wiechert potentials $A$ and $\phi$ by using (4.1) and (4.2). This task is very tedious but after a good deal of manipulation we find

$E(r, t)$

$$= \frac{q}{4\pi\varepsilon}\left[\frac{1}{R_0^3}\left(1 - \frac{u^2}{v^2}\right)\left(\boldsymbol{R} - \frac{\boldsymbol{u}R}{v}\right) + \frac{1}{v^2 R_0^3}\boldsymbol{R} \wedge \left\{\left(\boldsymbol{R} - \frac{\boldsymbol{u}R}{v}\right) \wedge \dot{\boldsymbol{u}}\right\}\right] \tag{4.72}$$

and

$$B = \frac{1}{Rv}(\boldsymbol{R} \wedge \boldsymbol{E}), \tag{4.73}$$

where $R_0 = R - \dfrac{\boldsymbol{R} \cdot \boldsymbol{u}}{v}$. The square brackets indicate that $\boldsymbol{u}$, $\dot{\boldsymbol{u}}$, $\boldsymbol{R}$ and $R$ must be evaluated at the retarded position of the charge for the point $r$ at time $t$. We shall comment on these results in the next chapter.

## PROBLEMS 4

1. Given potentials $A$ and $\phi$ that satisfy the Lorentz gauge condition and a Hertz potential $Z$ defined by

$$Z = \frac{1}{\mu\varepsilon}\int A\,dt + Z_0,$$

where $Z_0$ is a vector independent of $t$, show that div $Z = -\phi$

if div $Z_0 = -\phi_0$, where $\phi_0$ is the value of $\phi$ at $t = 0$. Hence, show that any given electric and magnetic fields $E$ and $H$ can be represented in terms of a suitably chosen vector $Z$.

2. Find a suitable Hertz potential to describe a plane wave of the form (2.16) and (2.17).

**CHAPTER 5**

# Sources of Electromagnetic Waves

## 5.1 Radiation from an accelerating point charge

There is a flow of electromagnetic energy from an accelerating charge, some of which is irretrievably lost to the system. We can measure this energy loss by the rate at which energy passes through a large sphere of radius $R$, centred at the retarded position, taken in the limit $R \to \infty$. This may be calculated from the surface integral of the normal component of the Poynting vector taken over the sphere

$$\int_0^{2\pi} \int_0^{\pi} (E \wedge H) \cdot RR \sin \theta d\theta d\psi, \tag{5.1}$$

taken in the limit $R \to \infty$. We shall refer to this quantity as the *radiation* loss, or simply the radiation.

The magnitude of the radiation depends upon the asymptotic form for the fields $E$ and $H$ at large distances from the charge. Only the terms in $E$ and $H$, which fall off as $1/R$ for large $R$, will contribute for (5.1) and, for this reason, these terms are often known as the radiation fields. We see from (4.71), the radiation fields for a point charge which has zero velocity ($[u] = 0$) at its retarded position are

$$E \sim \frac{q}{4\pi\varepsilon v^2} \left[ \frac{R \wedge (R \wedge \dot{u})}{R^3} \right], \tag{5.2}$$

and

$$H \sim \frac{q}{4\pi v} \left[ \frac{\dot{u} \wedge R}{R^2} \right]. \tag{5.3}$$

If the angle between $[\dot{u}]$ and $R$ is $\theta$, the power radiated per unit solid angle in the direction $\theta$, $\psi$ is

$$\frac{q^2 [\dot{u}]^2 \sin^2 \theta}{16\pi^2 \varepsilon v^3}. \tag{5.4}$$

The total rate of energy loss from the charge is

$$\frac{q^2 [\dot{u}]^2}{8\pi^2 \varepsilon v^3} \int_0^{\pi} \sin^3 \theta d\theta, \tag{5.5}$$

which is equal to $\mu q^2 [\dot{u}]^2 / 6\pi v$.

69

When the charge moves with a velocity $u$ the radiation fields become

$$E \sim \frac{q}{4\pi\varepsilon v^2}\left[\frac{R \wedge \{(R - uR/v) \wedge \dot{u}\}}{\left(R - \dfrac{u \cdot R}{v}\right)^3}\right] \tag{5.6}$$

and

$$H \sim \frac{q}{4\pi v}\left[\frac{(\dot{u} \wedge R)}{\left(R - \dfrac{R \cdot u}{v}\right)^3} + \frac{(u \wedge R)(\dot{u} \wedge R)}{v(R - u \cdot R/v)^3}\right]. \tag{5.7}$$

These equations are valid for $[u] < v$. The fields are zero if $[\dot{u}] = 0$, and hence there can be no radiation loss unless the charge is being accelerated or retarded.

The spherical waves that are emitted as the charge moves are not concentric, and this must be taken into account in the calculation of the energy loss. We must allow for the fact that the sphere of radius $R$ centred at the retarded position of the charge moves with a velocity $[u]$. A point $R$ on its surface travels with a speed $[u] \cdot R/R$ along the normal, so a spherical wave which diverges from the centre with a speed $v$ will pass through this point with a speed $v - [u] \cdot R/R$ along the outward normal. Thus the energy radiated per unit area through the sphere at a point $R$, per unit time, is the component of $E \wedge H$ along the normal, multiplied by the factor $\left[1 - \dfrac{u \cdot R}{vR}\right]$ to take into account the motion of the sphere.

When $u$ is in the same direction as the acceleration $\dot{u}$, the power radiated per unit solid angle is

$$\frac{q^2[\dot{u}]^2 \sin^2\theta}{16\pi^2\varepsilon v^3\left(1 - \dfrac{[u]}{v}\cos\theta\right)^5}. \tag{5.8}$$

The energy is radiated symmetrically about the direction of acceleration $[\dot{u}]$. For $[u] = 0$, the radiation is a maximum in the plane through the charge that is perpendicular to the direction of motion; it is also symmetrical about this plane. When $[u]$ is finite, the direction of maximum radiation is pushed *towards* the direction of $[u]$ but no energy is actually radiated along the line of $[u]$ (see Fig. 14).

70

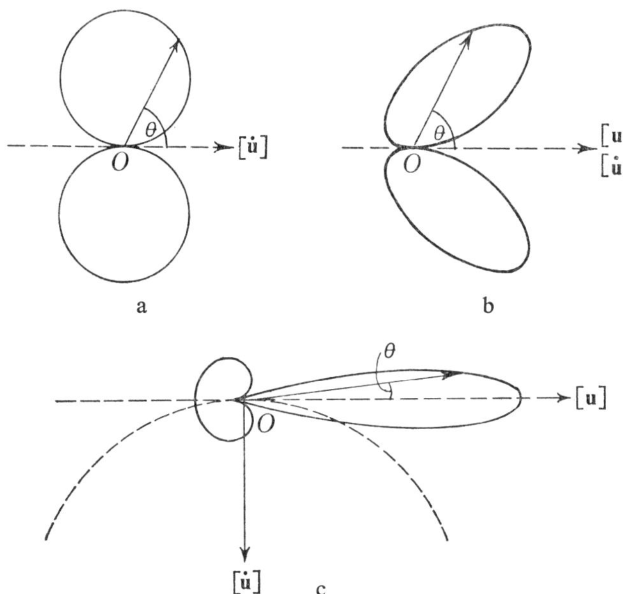



Fig. 14 The radiation from the charge at retarded positions $O$ in the direction $\theta$, $\psi$ (in polar coordinates, [u] as $z$-axis) is proportional to the length of the radius vector in this direction. (a) [u] $= 0$, (b) [u] and [u̇] in the same direction, (c) a charge moving in a circle. [u] and [u̇] are perpendicular

If the charge moves with a constant speed $a\omega$ in a circle of radius $a$, and we choose axes centred on the charge such that the $z$-direction is along $[\boldsymbol{u}]$ and tangential to the circle, and the $x$-direction is along $[\dot{\boldsymbol{u}}]$, the direction of the radius vector from the centre to the charge, then the energy radiated in the direction $\theta$, $\psi$ in polar coordinates is

$$\frac{q^2 a^2 \omega^4}{16\pi^2 \varepsilon v^3} \frac{\left\{\left(1 - \dfrac{a\omega}{v}\cos\theta\right)^2 - \left(1 - \dfrac{a^2\omega^2}{v^2}\right)\sin^2\theta\cos^2\psi\right\}}{\left(1 - \dfrac{a\omega}{v}\cos\theta\right)^5} \tag{5.9}$$

per unit angle.

The symmetry about the direction of motion has been lost. In this case energy is radiated along the instantaneous direction of motion and, when $a\omega$ is large (i.e. comparable with $v$) the radiation in this direction predominates.

The spectrum of the radiation field which will depend upon the time dependence of $[u]$ and $[\dot{u}]$, can be calculated by making a Fourier transformation of the electric field vector (5.6). In general the spectrum will be continuous and extend over all frequencies. Monochromatic waves can be generated by an oscillating charge. However, oscillating charges usually occur in pairs in the form of oscillating dipoles; we shall discuss the radiation from these in the next two sections.

The loss of energy of a charge due to radiation causes it to be retarded and contributes to its inertia. A complicated self-consistent equation results if we attempt to take this into account in the equation of motion. Fortunately, the force which causes this retardation is usually very small compared with the applied force and can be taken into account in a secondary way. This retarding force or 'radiation reaction' must do work at the rate at which energy is lost by radiation. If this force is $F$, and $u \ll v$, we can use the expression (5.5) for the rate of energy loss and equate it to the rate of working $F \cdot [u]$ giving

$$F \cdot [u] + \frac{q^2 \mu [\dot{u}]^2}{6\pi v} = 0. \tag{5.10}$$

This equation has the disadvantage that it cannot be solved explicitly in a general case to give an expression for $F$. If the equation is integrated over a time interval $t_1$ to $t_2$ such that the acceleration takes place completely within this interval, and $[\dot{u}] = 0$ at $t_1$ and $t_2$, we find

$$\int_{t_1}^{t_2} F \cdot [u] dt = \int_{t_1}^{t_2} \frac{\mu q^2}{6\pi v} [\ddot{u}] \cdot [u] dt, \tag{5.11}$$

where the integral on the right hand side has been re-expressed by an integration by parts. A force $F_R$, where $F_R = \mu q^2 [\ddot{u}]/6\pi v$, satisfies (5.11) when substituted for $F$. This force is usually taken as a measure of the radiation reaction because it is more convenient than the solution of (5.10).

If one calculates the effect of the fields of the charge itself in an attempt to describe in detail the mechanism of the radiation reaction, fundamental difficulties occur. The self-energy associated with this self-force becomes very large for a small charge and infinite for a point charge. This is a limitation of the classical approach to electrodynamics, but similar problems occur in quantum electrodynamics. A detailed discussion of this is beyond the scope of this book, but we refer the reader who is interested to [4] in the bibliography.

## 5.2 Electric dipole radiation

For a simple model of a time dependent dipole, we consider two charges $q(t)$ and $-q(t)$ at points $(0, 0, a)$ and $(0, 0, -a)$ respectively, and joined by a short wire along the $z$-axis. This system has an electric dipole moment in the $z$-direction of magnitude $2aq(t)$.

The charge density for the system can be described by the distribution function

$$\rho(\mathbf{r}, t) = q(t)\delta(x)\delta(y)\{\delta(z-a)-\delta(z+a)\}, \tag{5.12}$$

where $\delta(x)$ is the Dirac delta-function discussed in section (4.3).

As a first step in the calculation of the radiation from this dipole system, we shall calculate the potentials $A$ and $\phi$. If we substitute (5.12) into the expression (4.31) for the retarded scalar potential, and use the property (4.50) of the $\delta$-function, we find

$$\phi(\mathbf{r}, t) = \frac{1}{4\pi\varepsilon}\left\{\frac{q\left(t-\dfrac{R_+}{v}\right)}{R_+} - \frac{q\left(t-\dfrac{R_-}{v}\right)}{R_-}\right\}, \tag{5.13}$$

where

$$R_\pm^2 = r^2+a^2\mp2ar\cos\theta, \tag{5.14}$$

in terms of the spherical coordinates $(r, \theta, \psi)$.

If the distance is very small compared with the distance to the point of observation $\mathbf{r}$, and the velocity of the charge flow along the wire is small compared with $v$, we need only consider the leading terms in a Taylor's series expansion of (5.13) in powers of $a$. The expansion of one of the terms appearing in (5.13) is

$$\frac{q\left(t-\dfrac{R_+}{v}\right)}{R_+} = \frac{q\left(t-\dfrac{r}{v}\right)}{r}$$

$$+ \frac{a}{1!}\left(\left\{\frac{-q\left(t-\dfrac{r}{v}\right)}{r^2} - \frac{\dot{q}\left(t-\dfrac{r}{v}\right)}{vr}\right\}\frac{dR_+}{da}\right)_{a=0} +..., \tag{5.15}$$

where we have used the relation

$$\frac{d}{da}q\left(t-\frac{R_+}{v}\right) = -\frac{1}{v}\frac{d}{dt}q\left(t-\frac{R_+}{v}\right)\frac{dR_+}{da}. \tag{5.16}$$

From (5.14) we find that $\dfrac{dR_+}{da}$ evaluated at $a = 0$ is $-\cos\theta$ so that, to leading order in $a$, the scalar potential is given by

$$\phi(r, t) \sim \frac{2a\cos\theta}{4\pi\varepsilon r}\left[\frac{q}{r} + \frac{\dot{q}}{v}\right]. \tag{5.17}$$

To calculate the corresponding vector potential we require an expression for the current distribution. If the wire has negligible thickness the current density is given by

$$j(r, t) = k_0 \frac{dq}{dt}\,\delta(x)\delta(y)\theta(z+a)(1-\theta(z-a)), \tag{5.18}$$

where $\theta(z)$ is the unit step function defined by

$$\theta(z) = \begin{cases} 0, & z<0, \\ 1, & z\geqq 0, \end{cases} \tag{5.19}$$

and $k_0$ is a unit vector in the $z$-direction.

If we substitute this form for the current density into (4.32) for the vector potential, we find, to leading order in $a$,

$$A(r, t) \sim \frac{2\mu a}{4\pi}\frac{\dot{q}}{r}\,k_0. \tag{5.20}$$

We can introduce the concept of a point dipole by allowing the dimensions of the system tend to zero in such a way that $\underset{a\to 0}{\text{limit}}\,aq(t)$ remains finite. If $p(r, t)$ is the dipole distribution function, in this limit we obtain a point dipole situated at the origin such that

$$p(r, t) = p_0(t)\delta(x)\delta(y)\delta(z), \tag{5.21}$$

where $p_0(t) = 2k_0\,\underset{a\to 0}{\text{limit}}\,aq(t)$ in this model. The potentials (5.17) and (5.20) in this limit are given by

$$\phi(r, t) = \frac{\cos\theta}{4\pi\varepsilon r}\left[\frac{p_0}{r} + \frac{\dot{p}_0}{v}\right], \tag{5.22}$$

and

$$A(r, t) = \frac{\mu}{4\pi r}\,\dot{p}_0. \tag{5.23}$$

We shall assume that these results are independent of the particular model of a dipole we have chosen. The form of (5.22) and (5.23) can be checked for another model of an oscillating dipole which is given in problem one at the end of the chapter.

The fields $E$ and $H$ can be calculated from the potentials (5.22) (5.23) by making a substitution into (4.1) and (4.3). It is quite straightforward though a little tedious to show that they can be written in the form

$$E = \frac{1}{4\pi\varepsilon r^3}\left[\frac{3r(r\cdot p_0)}{r^2} - p_0\right] + \frac{1}{4\pi\varepsilon v r^2}\left[\frac{3r(r\cdot \dot{p}_0)}{r^2} - \dot{p}_0\right]$$

$$+ \frac{1}{4\pi\varepsilon v^2 r^3}[r\wedge(r\wedge \ddot{p}_0)], \qquad (5.24)$$

and

$$H = -\frac{1}{4\pi r}\left[\frac{r\wedge \ddot{p}_0}{v} + \frac{r\wedge \dot{p}_0}{r}\right]. \qquad (5.25)$$

The magnetic field $H$ is perpendicular to the plane containing the radius vector $r$ and the dipole $p$. For $\dot{p}_0 = \ddot{p}_0 = 0$, (5.24) and (5.25) reduce to the usual expression for the fields associated with a small static dipole. When $\ddot{p}_0 \neq 0$ the dipole generates spherical waves. The asymptotic form of the fields for large $r$ from (5.24) and (5.25) are

$$E \sim \frac{\mu}{4\pi}\left[\frac{r\wedge(r\wedge \ddot{p}_0)}{r^3}\right] \qquad (5.26)$$

and

$$H \sim \frac{1}{4\pi v}\left[\frac{\ddot{p}_0 \wedge r}{r}\right]. \qquad (5.27)$$

These fields are perpendicular and give rise to a Poynting vector which lies along the radius vector $r$. The rate at which energy is radiated is calculated by substituting (5.26) and (5.27) into (5.1) giving

$$\int_{\substack{\text{limit} \\ R\to\infty}} (E\wedge H)\cdot RR \sin\theta d\theta d\psi = \frac{\mu[\ddot{p}_0]^2}{8\pi v}\int_0^\pi \sin^3\theta d\theta, \qquad (5.28)$$

which is equal to $\dfrac{\mu[\ddot{p}_0]^2}{6\pi v}$.

An alternative to the calculation of $A$ and $\phi$ would be to calculate the Hertz vector $Z(r, t)$ directly from the dipole distribution function

(5.21). The retarded solution of (4.10) is of the same form as (4.32). For a point dipole the integration is simple, and we find a retarded potential,

$$Z(r, t) = \frac{1}{4\pi\varepsilon} \frac{p_0\left(t - \dfrac{r}{v}\right)}{r}. \tag{5.29}$$

The expressions (5.24) and (5.25) are obtained for $E$ and $H$ on substituting (5.29) into (4.13) (4.14) and carrying out the differentiation.

### 5.3 Magnetic dipole radiation

The fields associated with a small magnetic dipole can be calculated in a similar way to those for the electric dipole considered in the previous section.

For a model of a small magnetic dipole we consider a current $I(t)$ flowing in a small circular wire of radius $a$. This system has a total magnetic dipole moment of magnitude $\pi a^2 I(t)$ which is in a direction perpendicular to the plane of the circle. Let the circle be the $xy$-plane and be centred at the origin. The contribution to the retarded vector potential due to a current flowing in a small element of length $a\delta\psi_1$, is

$$\frac{\mu}{4\pi} a\delta\psi_1 \frac{I\left(t - \dfrac{R_1}{v}\right)}{R_1} \tag{5.30}$$

in the direction of the current flow, where $R_1$ is given by

$$R_1^2 = r^2 + a^2 - 2ar \sin\theta \cos\psi_1. \tag{5.31}$$

The angle $\psi_1$ is measured in the $xy$-plane with respect to the projection of $r$ on this plane (see Fig. 15).

If this element is combined with an element in the direction $-\psi_1$ the combined contribution to $A$ at the point $P$ is in the $\psi$ direction. The total contribution is obtained by an integration with respect to $\psi_1$,

$$A_\psi = \frac{2a\mu}{4\pi} \int_0^\pi \frac{I\left(t - \dfrac{R_1}{v}\right)}{R_1} \cos\psi_1 d\psi_1. \tag{5.32}$$

For small values of $a$ the integrand can be expanded in a Taylor's

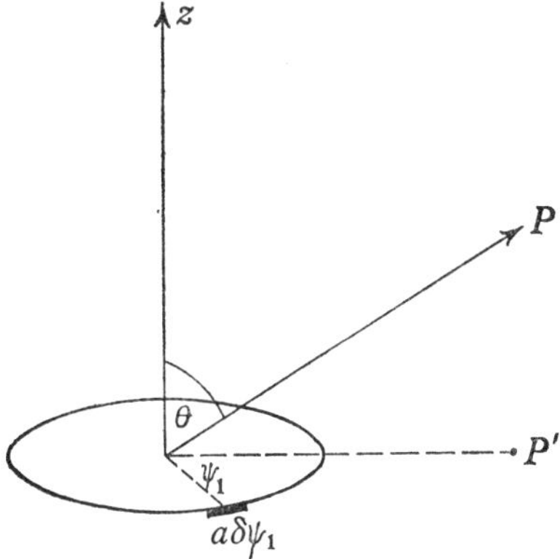

Fig. 15 A current flow in a small circular wire of radius $a$. $P$ is the observation point $(r, \theta\psi)$, and $OP'$ is the projection of $OP$ on to the $xy$-plane

Series in powers of $a$ as follows

$$\frac{I\left(t-\frac{R_1}{v}\right)}{R_1} = \frac{I\left(t-\frac{r}{v}\right)}{r} - \frac{a}{1!}\left[\frac{1}{r^2} + \frac{\dot{I}}{vr}\right]\frac{\partial R_1}{\partial a}\bigg)_{a=0} + \dots \tag{5.33}$$

From (5.31) $\dfrac{dR_1}{da} = \sin\theta\cos\psi_1$ at $a = 0$.

To leading order in $a$ the vector potential is

$$A_\psi \sim \frac{a^2\mu}{2\pi r}\int_0^\pi \left[\frac{I}{r} + \frac{\dot{I}}{v}\right]\cos^2\psi_1 d\psi_1. \tag{5.34}$$

On performing the integration, we find

$$A_\psi \sim \frac{\mu}{4\pi r}\left[\frac{m_0}{r} + \frac{\dot{m}_0}{v}\right]\sin\theta, \tag{5.35}$$

where we have used the notation $m_0 = \pi a^2 I k_0$.

We can use an equality in (5.35) in the limit $a\to 0$ such that $\lim\limits_{a\to 0} \pi a^2 I = m_0$. Equation (5.35) then gives the vector potential, due to a point magnetic dipole. The other components $A_r$ and $A_\theta$ are zero and so is the scalar potential $\phi$.

The fields can be calculated from the vector potentials in the usual way. The results are very similar to those of an electric dipole and can be obtained from (5.24) and (5.25) by the transformation $p_0 \rightarrow m_0$, $E \rightarrow -H$ and $H \rightarrow E$. It follows directly from this that the rate at which a magnetic dipole radiates energy is $\dfrac{\mu[\ddot{m}_0]^2}{6\pi v}$.

An electric or magnetic dipole which oscillates at a single frequency $\omega$ emits monochromatic waves of wavelength $2\pi v/\omega$; if the dipole moment strengths, $p_0(t)$ and $m_0(t)$, vary as $e^{i\omega t}$ and the fields have the same form of time dependence. For an oscillating electric or magnetic dipole distribution of macroscopic dimensions, formulae (5.24) (5.25) still give a good description of the radiation fields provided the wavelength of the waves emitted is large compared to the extent of the dipole distribution. In this case $p_0$ and $m_0$ in (5.24) (5.25) refer to the electric or magnetic dipole moment of the total distribution.

Waves in the visible part of the electromagnetic spectrum are generated from single atoms due to the atomic electric dipole moment. The spectrum of the radiation can be calculated if one uses some specific model of an atom. In a semi-classical model the electrons are assumed to obey classical mechanics and oscillate about the nucleus due to an harmonic force, together with a small amount of damping due to the radiation reaction. Such a model is unstable because the electrons lose energy continuously and eventually fall into the nucleus. However, some of the predictions based on this model are qualitatively correct. For a simple explanation of dispersion see Question 3, Chapter 2, and for scattering see Question 2, Chapter 6.

In a quantum mechanical description of the atom the electrons exist in stable atomic shells about the nucleus. Radiation is emitted only when an electron makes a transition from one shell to another. The frequency of the light emitted is directly proportional to the energy difference between the two shells involved in the transition. Consequently, the emitted spectrum consists of a number of narrow lines, whose frequencies reflect the structure of the electronic states of the atom. This description, however, goes beyond the macroscopic framework within which we have been working. To apply formulae (5.24) (5.25) the dipole source must be small compared with the wavelength of the radiation, but to be consistent with our original assumptions, it must be large on atomic scale even when we refer to it as a 'point dipole'.

## 5.4 Radiation from an antenna

When the extent of the dipole source is not small as compared with the wavelength of the radiation the calculation of the retarded potential is rather more difficult. An example of this type of situation is a current-carrying wire or antenna which emits radio waves of wavelength comparable with the length of the wire. Consider an antenna of length $2l$ situated at the origin and pointing in the $z$-direction. If it carries a current $I = I_0(z')e^{i\omega t}$ at a point $z'$ along its length, it will be equivalent to an electric dipole distribution

$$p(r', t) = k_0 \frac{I_0(z')}{i\omega} e^{i\omega t}\delta(x')\delta(y'), \qquad -l \leqq z' \leqq l \tag{5.36}$$

which was found by using equation (4.11). The retarded Hertz potential corresponding to this distribution is

$$Z(r, t) = \frac{k_0}{4\pi i\omega\varepsilon} \int_{-l}^{l} \frac{I_0(z')e^{i\omega\left(t - \frac{|r-k_0 z'|}{v}\right)}}{|r-k_0 z'|}\, dz'. \tag{5.37}$$

As it is our intention to calculate the radiation we need only the asymptotic form of the fields for large $r$ and, hence, only the asymptotic form for $Z(r, t)$. By making the approximation

$$|r-k_0 z'| \sim r - z' \cos\theta + O\left(\frac{1}{r}\right), \tag{5.38}$$

we can show that the asymptotic form for $Z$ is given by

$$Z(r, t) = \frac{k_0 e^{i\omega\left(t - \frac{r}{v}\right)}}{4\pi i\omega\varepsilon r} \int_{-l}^{l} I_0(z')e^{\frac{i\omega z' \cos\theta}{v}}\, dz'. \tag{5.39}$$

The potential, and hence the fields, can be calculated once the current distribution $I_0(z')$ is known. This is really a very difficult self-consistent problem because $I_0(z')$ will be affected to some degree by the radiation. However, it is usually a reasonable approximation to take $I_0(z')$ as the applied current distribution in the wire.

Let us consider the simple case of a half-wave antenna. The wire carries a standing wave which has a wavelength equal to twice the length of the wire. Hence, the current distribution is

$$I_0(z') = I_0 \cos\left(\frac{z'\pi}{2l}\right). \tag{5.40}$$

If this is substituted into (5.39), the integral is quite easy to evaluate, and we obtain the result

$$Z(r,\,t) = \frac{k_0 e^{i\omega\left(t-\frac{r}{v}\right)}}{4\pi i \omega^2 \varepsilon r}\, v I_0 \left\{ \frac{\sin\left(\frac{\pi}{2}(1+\cos\theta)\right)}{(1\pm\cos\theta)} - \frac{\sin\left(\frac{\pi}{2}(1-\cos\theta)\right)}{(1-\cos\theta)} \right\}.$$

$$(5.41)$$

This can be simplified to

$$Z(r,\,t) = \frac{k_0 e^{i\omega\left(t-\frac{r}{v}\right)}}{2\pi i \omega^2 \varepsilon r}\, v I_0 \, \frac{\cos\left(\frac{\pi}{2}\cos\theta\right)}{\sin^2\theta}. \qquad (5.42)$$

The asymptotic form of the fields can be calculated by substituting (5.42) in (4.13); the results are

$$E_\psi \sim \frac{I_0 \mu i \cos\left(\frac{\pi}{2}\cos\theta\right)}{2\pi r \sin\theta}\, e^{i\omega\left(t-\frac{r}{v}\right)}, \qquad (5.43)$$

$$H_\theta \sim \frac{I_0 i \cos\left(\frac{\pi}{2}\cos\theta\right)}{2\pi r \sin\theta}\, e^{i\omega\left(t-\frac{r}{v}\right)}. \qquad (5.44)$$

With this particular current distribution the radiation has similar features to that from a point dipole. There is no radiation along the length of the antenna and the maximum is in the plane $\theta = \pi/2$.

The factors which determine the distribution of the radiation from an array of antennae are the form of the current distribution $I(z')$, the shape of the individual antenna, and their relative positions. The problems of antenna design amount to finding an optimum form of these factors for the efficient propagation of waves to the receiving or detecting devices.

### 5.5 Radiation from an arbitrary charge and current distribution

Consider a finite charge and current distribution which oscillates harmonically in time with a frequency $\omega$. Let the charge and current densities be of the form $\rho_0(r')e^{i\omega t}$ and $j_0(r')e^{i\omega t}$ respectively. These will give rise to retarded potentials $A$ and given by

$$\phi(r,\,t) = \frac{1}{4\pi\varepsilon} \int \frac{\rho_0(r')e^{i\omega\left(t-\frac{|r-r'|}{v}\right)}}{|r-r'|}\, d\tau', \qquad (5.45)$$

and

$$A(r, t) = \frac{\mu}{4\pi} \int \frac{j_0(r')e^{i\omega\left(t - \frac{|r-r'|}{v}\right)}}{|r-r'|}\, d\tau', \qquad (5.46)$$

in a dielectric characterized by the constants $\varepsilon$ and $\mu$. The origin of the vector $r$ and $r'$ is assumed to be situated at some point within the distribution.

If we are interested in calculating the radiation from this distribution we need only consider the leading term in an asymptotic expansion of the fields for large $r$. The factor in the integrand which depends upon $r$ behaves asymptotically as

$$\frac{e^{-i\frac{\omega}{v}|r-r'|}}{|r-r'|} \sim \frac{e^{-i\frac{\omega r}{v}}}{r}\, e^{i\frac{\omega r'}{v}\cos\theta'}, \qquad (5.47)$$

where $\theta'$ is the angle between the vectors $r$ and $r'$. On substituting (5.47) into (5.45) and (5.46) we find

$$\phi(r, t) \sim \frac{e^{i(\omega t - kr)}}{4\pi\varepsilon r} \int \rho_0(r')e^{ikr'\cos\theta'}\, d\tau', \qquad (5.48)$$

and

$$A(r, t) \sim \frac{\mu e^{i(\omega t - kr)}}{4\pi r} \int j_0(r')e^{ikr'\cos\theta'}\, d\tau', \qquad (5.49)$$

where $k = \omega/v$. If we denote these asymptotic forms, (5.48) and (5.49), by $\phi_{\text{rad}}$ and $A_{\text{rad}}$ respectively, and substitute into (4.1) and (4.3) to calculate the radiation fields, we find

$$B_{\text{rad}} = -\frac{ikr \wedge A_{\text{rad}}}{r} \qquad (5.50)$$

and

$$E_{\text{rad}} = \frac{ik}{r}\, \phi_{\text{rad}}\, r - i\omega A_{\text{rad}}, \qquad (5.51)$$

which are exact results for the *radiation* fields.

When the dimensions of the source are small compared to the wavelength of the radiation ($kr' \ll 1$) it may be reasonable to expand the exponential term $e^{ikr'\cos\theta'}$ in the integrals in (5.48) and (5.49) in powers of $kr'$. The integral of the first term in (5.48), if this expansion is made, will be the total charge of the system. The corresponding radiation is known as monopole radiation. The second term will be the electric dipole moment of the distribution in the direction of $r$, and the associated radiation is known as dipole radiation. The third term is the integral of $k^2 r'^2 \cos^2 \theta'$ over the distribution $\rho_0(r')$. This is the first non-vanishing term for a simple quadrupole system but its value is not the quadrupole moment. A simple model of a quadrupole is a system of charges, $2q$ at $(0, 0, 0)$ and two charges of $-q$ at $(0, 0, a)$ and $(0, 0, -a)$. This is a linear quadrupole and is essentially two opposing dipoles displaced from one another along the same straight line. The quadrupole moment of the general charge distribution $\rho_0(r')$ in the direction of $r$ is

$$\int \rho_0(r')\left(\frac{3\cos^2\theta'-1}{2}\right)d\tau'. \tag{5.52}$$

The $\theta'$ dependence can be written as $P_2(\cos\theta')$, similarly the higher moments are integrals over higher order Legendre polynomials $P_l(\cos\theta')$. The intensity of the quadrupole radiation is weaker than the electric dipole radiation by a factor $k^2$.

In the expression (5.49) for $A(r, t)$ the zero order term in $k$ is non-vanishing for electric dipole radiation, and the first order term in $k$ is associated with magnetic dipole radiation.

## 5.6 Cerenkov radiation

So far we have given the impression that only an accelerating charge can radiate. This is not strictly true. A charge travelling with a constant velocity in a dielectric medium will radiate if its velocity exceeds the velocity of light in the dielectric. The charge passes through a 'light barrier' and, as a consequence, it produces a shock wave. The effect was first observed by Cerenkov and is known as Cerenkov radiation.

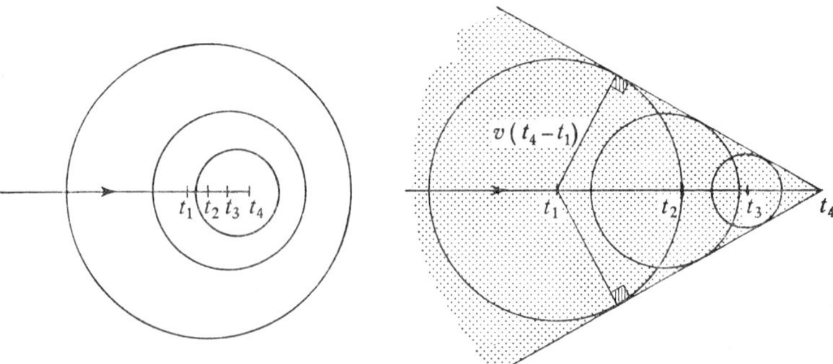


Fig. 16 Spherical light waves from a charge travelling with uniform velocity emitted at times $t_1$, $t_2$, $t_3$ and $t_4$, $(t_4 > t_3 > t_2 > t_1)$

When $u<v$ the spherical waves that are emitted from a charge at consecutive times $t_1$, $t_2$ $t_3$, ... do not overlap and each spherical wave completely contains those that are emitted at later times. For $u>v$, however, the spherical waves intersect in such a way that their envelope forms a Mach cone of semi-angle $\sin^{-1}\left(\dfrac{v}{u}\right)$. The equation of the cone is

$$(x-ut)^2+(y^2+z^2)\left(1-\frac{u^2}{v^2}\right) = 0 \qquad (5.53)$$

for a charge that travels in the $x$-direction. Outside the cone the electric and magnetic fields are zero. Inside the cone the fields are due to two waves which have originated from different retarded positions. In Fig. 17 we have attempted to redraw Fig. 12 in the case $u>v$, showing two retarded positions $A'$ and $A''$. There are two points $D'$ and $D''$ which correspond to the point $D$ in Fig. 13. If we refer back to equation (4.53) and consider the case $u>v$ we see that there are two possible solutions for PD, one positive and the other negative. Because PD is equal to $R-\dfrac{\boldsymbol{R}\cdot\boldsymbol{u}}{v}$, and this must always be positive for $u<v$, we can only take one solution for PD. However, for $u>v$

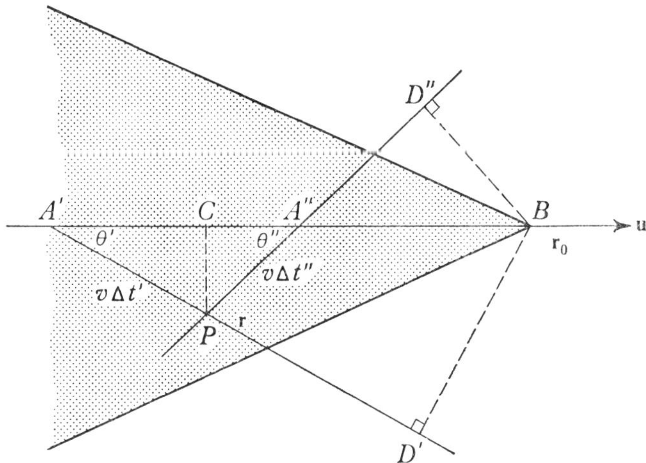


Fig. 17 This is Figure 13 redrawn for the cose $u>v$. $A'$ and $A''$ are the two retarded positions for the observation point $P$ which lies with the Mach cone. When the charge travels with a constant velocity $\boldsymbol{u}$, $A'B = \boldsymbol{u}$ and $A''B = \boldsymbol{u}\,t''$.

the quantity $R - \dfrac{R \cdot u}{v}$ can have positive and negative values, and both solutions of (4.53) are valid. They correspond to the two values $PD'$ and $PD''$.

From equation (4.43) we see that $\left| R - \dfrac{R \cdot u}{v} \right|$ is required because we must take the positive value for the Jacobian of the transformation. The modulus of the two solutions of (4.53) are equal, so the potentials at point within the cone are twice those given by (4.58) and (4.59). Hence, the field values within the cone are

$$E(r, t) = \frac{q}{2\pi\varepsilon} \frac{(r - r_0)\left(1 - \dfrac{u^2}{v^2}\right)}{\left\{(x - ut)^2 + (y^2 + z^2)\left(1 - \dfrac{u^2}{v^2}\right)\right\}^{\frac{3}{2}}} \qquad (5.54)$$

and

$$H(r, t) = \frac{\mu q}{2\pi} \frac{u \wedge (r - r_0)\left(1 - \dfrac{u^2}{v^2}\right)}{\left\{(x - ut)^2 + (y^2 + z^2)\left(1 - \dfrac{u^2}{v^2}\right)\right\}^{\frac{3}{2}}}. \qquad (5.55)$$

At any point within the cone the electric vector is directed towards the present position of the charge. Therefore, at points almost on the surface of the cone the electric field lies along the cone generator. The magnetic field at these points is tangential to the cross-section of the cone perpendicular to $u$, and so the Poynting vector is normal to the surface. At points actually on the cone the field values are singular. This is due to the singular nature of a point charge; a finite charge gives fields that pass through a maximum in the region of the cone and the singularity is smoothed out.

The radiation is essentially in the form of a shock wave because only the field values near the edge of the cone area are significant. The total rate of radiation loss can be calculated even for a point charge but the singularities in the field values cause certain formal difficulties and we refer the reader to the paper by Tamm, who gave the original explanation of the phenomenon (see [5]). It can be proved that in a more realistic calculation in which dispersion is taken into account, the shock wave travels with the group velocity rather than the phase velocity $v$, and that the radiation spectrum lies mainly in the visible frequency range.

## PROBLEMS 5

1. Two charges $q$ and $-q$ are at points $(0, 0, af(t))$ and $(0, 0, -af(t))$ respectively, where $f(t)$ is a function of time. Use the Liénard Wiechert potentials to find the vector and scalar potentials at a point $(x, y, z)$. Show that in the limit $a \to 0$ such that $2aqf(t) \to p_0$ the potentials are given by (5.22) and (5.23).

2. Calculate the total rate of radiation loss from a point charge $q$ that moves with constant angular velocity $\omega$ round a circle of radius $a$.

3. The $x$-coordinate of an electron of charge $e$ and mass $m$ satisfies the equation

$$\frac{d^2x}{dt^2} + \gamma \frac{dx}{dt} + \omega_0^2 x = \frac{eE_0}{m} e^{i\omega t}$$

in the presence of an electric field $E_0 e^{i\omega t}$. Show that the electron has a dipole moment associated with the forced oscillations of magnitude,

$$\frac{E_0 e^2}{m} \frac{e^{i\omega t}}{\omega_0^2 - \omega^2 + i\omega\gamma}.$$

Hence, calculate the total rate of dipole radiation due to the forced oscillations. Show that it is proportional to $\omega^4$ if $\omega \ll \omega_0$.

4. A linear quadrupole oscillator consists of a charge $2q$ at the origin, and two charges each of $-q$ at the points $(0, 0, a \sin \omega t)$ and $(0, 0, -a \sin \omega t)$. Find the average rate at which energy is radiated.

CHAPTER 6

# Scattering and Diffraction

## 6.1 Introduction

Scattering is the term used to describe the modification of a beam of electromagnetic waves caused by an obstacle in its path. When the wavelength $\lambda$ of the incident waves is very much smaller than the dimensions of the obstacle, a region of shadow is created, which is only well-defined in the limit of geometrical optics ($\lambda \to 0$), and in this limit corresponds to the region of zero intensity. The modulation in intensity and the scattering of waves into the region of shadow for finite but small wavelengths is known as *diffraction.*

The mathematical description of diffraction, and the more general problem of scattering is extremely difficult, and few problems can be solved exactly. In those that can, some advanced mathematical techniques are required. For most problems one has to be content with some form of approximate solution. Types of approximations vary but generally fall into one of the three categories (*a*) an approximation valid for some range of parameter, or parameters, involved in the problem (and might be asymptotically exact in some cases); (*b*) an approximation based on some numerical solution of the equations; (*c*) a simplification based on some intuitive insight into the physical nature of the problem. The accuracy of approximations of the type (*c*) is often not known a priori and depends upon a postiori experimental justification.

The scattering of waves caused by an obstacle without edges is usually approached as a typical boundary value problem. The incident wave plus the scattered outgoing wave are matched to the solution of Maxwell's equations valid within the obstacle, by applying the usual boundary conditions at the interface. For some purposes it is convenient to express the results in the form of a *differential scattering cross-section* $\sigma(\theta, \psi)$. This is a measure of the intensity of the scattered wave in the direction with polar coordinates $\theta$, $\psi$ (the $z$-direction being the direction of incidence) at large distances from the obstacle. It is defined as the asymptotic rate of energy

86

flow of the scattered wave per unit solid angle in the direction $\theta$, $\psi$, divided by the intensity of the incident wave. As an illustration of the method of calculation for this type of scattering, we shall consider a fairly simple example of a plane wave scattered by an infinite, perfectly conducting cylinder.

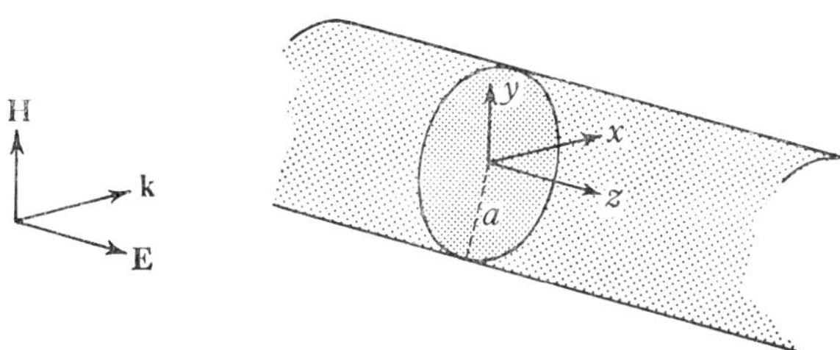


Fig. 18 A plane polarized wave incident in the $x$-direction on a perfectly conducting circular cylinder. The electric vector of the wave is parallel to the axis of the cylinder

## 6.2 Scattering by a perfectly conducting cylinder

Let the cylinder have a radius $a$ and let its axis be in the $z$-direction. If a plane polarized wave is incident in the $x$-direction with its electric vector parallel to the $z$-axis, this vector $E_z$ will be of the form,

$$E_z = E_0 e^{i(\omega t - kr \cos \theta)}, \tag{6.1}$$

in cylindrical coordinates $(r, \theta, z)$ where $k = \omega\sqrt{\varepsilon\mu}$ and $\varepsilon$, $\mu$ characterize the dielectric medium surrounding the cylinder.

The wave scattered by the cylinder must be an outgoing cylindrical wave independent of the $z$-coordinate. Its electric vector must be in the $z$-direction and must satisfy the *scalar* wave equation at all points outside the cylinder

$$\nabla^2 E_z = \varepsilon\mu \frac{\partial^2 E_z}{\partial t^2}, \tag{6.2}$$

because $E_x$ and $E_y$ are zero. We can find product solutions of this equation in cylindrical coordinates by substituting $E_z = R(r)\Theta(\theta)$, and then performing the usual method of separation of variables to obtain independent equations for $R$ and $\Theta$. The function $\Theta$ must satisfy the equation,

$$\frac{d^2\Theta}{d\theta^2} + n^2\Theta = 0, \tag{6.3}$$

G

where the separation constant $n^2$ has been chosen for convenience so that solutions can be written as $e^{\pm in\theta}$. The function $R$ must satisfy the equation

$$\frac{d^2R}{dr^2} + \frac{1}{r}\frac{dR}{dr} + \left(k^2 - \frac{n^2}{r^2}\right)R = 0, \tag{6.4}$$

which is Bessel's equation of order $n$ in terms of the variable $kr$. The solutions are $J_n(kr)$ and $Y_n(kr)$, Bessel functions of the first and second kind. The electric vector of the scattered wave can be constructed from a linear superposition of those solutions which give an outgoing wave

$$E_z = \sum_{n=-\infty}^{\infty} a_n\{J_n(kr) - iY_n(kr)\}e^{i(\omega t + n\theta)}. \tag{6.5}$$

The constant $n$ must be an integer so that the term $e^{+in\theta}$, and hence $E_z$, is single-valued.

The combination of Bessel functions $J_n(kr) - iY_n(kr)$ is the only one which results in a completely outgoing wave.

For a perfectly conducting cylinder there can be no penetration of the wave into the cylinder and the boundary condition at $r = a$ is simply that the total tangential electric vector should vanish for all $z$ and $t$. Hence

$$E_{0z}e^{i(\omega t - ka\cos\theta)} + \sum_{n=-\infty}^{\infty} a_n\{J_n(ka) - iY_n(ka)\}e^{i(\omega t + n\theta)} = 0. \tag{6.6}$$

To find the coefficients $a_n$ which satisfy this equation we expand $e^{-ika\cos\theta}$ in a complex Fourier series as

$$e^{-ika\cos\theta} = \sum_{n=-\infty}^{\infty} b_n e^{in\theta}, \tag{6.7}$$

where the coefficients $b_n$ are given by

$$b_n = \frac{1}{2\pi}\int_0^{2\pi} e^{-i(n\theta + ka\cos\theta)}d\theta. \tag{6.8}$$

This integral corresponds to an integral representation of the Bessel function of the first kind of order $n$, apart from a factor $(-i)^n$, so we can write $b_n = (-i)^n J_n(ka)$. This step is really equivalent to expanding a plane wave as a linear combination of cylindrical waves. If we substitute (6.7) and (6.8) into (6.6) we can equate each coefficient of the functions $e^{in\theta}$ equal to zero because they are linearly independent. This results in the following expression for $a_n$

$$a_n = - \frac{(-i)^n E_{0z} J_n(ka)}{J_n(ka) - i Y_n(ka)},\tag{6.9}$$

from which we can find the form of the scattered wave. The magnetic vector associated with the wave can be found by substituting $E_z$ into Maxwell's equation (1.4) in cylindrical coordinates

$$H_\theta = - \frac{i}{\omega\mu} \frac{\partial E_z}{\partial r}.\tag{6.10}$$

The other two components $H_z$ and $H_r$ are zero.

Though we have solved the problem exactly, the expression for the scattered wave is in the form of an infinite series which does not always rapidly converge. For this reason it is particularly difficult to find the form of the diffraction pattern when relatively short wavelength $(ka \gg 1)$ waves are scattered, and to examine the behaviour of the complete solution in the region of shadow behind the cylinder.

At large distances from the cylinder $(r \gg a)$ the scattered wave behaves asymptotically as

$$E_z \sim \sum_{n=-\infty}^{\infty} a_n \sqrt{\frac{2}{\pi k r}}\, e^{i\left(\omega t - kr - \frac{n\pi}{2} - \frac{\pi}{4}\right) + in\theta}\tag{6.11}$$

and

$$H_\theta \sim \sqrt{\frac{\varepsilon}{\mu}}\, E_z.$$

The associated Poynting vector $S$ for large $r$ lies in the $r$-direction and its time average is given by

$$S_r = \tfrac{1}{2} Re(E \wedge H^*) \sim \tfrac{1}{2} \sqrt{\frac{\varepsilon}{\mu}}\, |E_z|^2,\tag{6.12}$$

where $E_z$ is given by (6.11). The intensity of the scattered wave in the $\theta$-direction per unit length of cylinder and per unit angle is $rS_r$. The ratio of this quantity to the intensity of the incident wave, $\tfrac{1}{2}\sqrt{\varepsilon/\mu}\,|E_{0z}|^2$, in the limit $r \to \infty$ is the differential scattering cross-section $\sigma(\theta)$. From equation (6.11) and (6.12) we find for $\sigma(\theta)$

$$\sigma(\theta) = \frac{2}{\pi k} \left| \sum_{n=-\infty}^{\infty} \frac{J_n(ka)(-i)^n e^{in\theta}}{J_n(ka) - i Y_n(ka)} \right|^2.\tag{6.13}$$

This is a very complicated double series due to interference between the different cylindrical scattered waves, which are often known as *partial waves*. For the total rate of relative scattering we integrate (6.13) with respect to $\theta$ over the range 0 to $2\pi$. Due to the orthogonality properties of the functions $e^{in\theta}$ the double summation reduces to a single one giving

$$\sigma = \int_0^{2\pi} \sigma(\theta)d\theta = \frac{4}{\pi k} \sum_{n=-\infty}^{\infty} \left| \frac{J_n^2(ka)}{J_n^2(ka) + Y_n^2(ka)} \right|. \tag{6.14}$$

The partial waves contribute independently to the total scattering cross-section $\sigma$. The total intensity of the scattered wave is the sum of the intensities of the partial waves. Only in the very long wave limit $ka \ll 1$ is the scattering dominated by the first one or two partial waves. From the asymptotic form of the Bessel functions in the limit $ka \to 0$ we can show that the leading term in the asymptotic form for $\sigma$ comes from the partial wave $n = 0$, and is

$$\sigma \sim \frac{\pi^2}{k \log^2 ka}. \tag{6.15}$$

When the incident plane wave is polarized such that its *magnetic* vector lies in the $z$-direction, we can calculate the scattering in a similar way. The only point at which the calculations differ is in applying the boundary condition at $r = a$. At the surface of a perfect conductor the normal derivative of the magnetic field must vanish, namely

$$\left( \frac{\partial H_z}{\partial r} \right)_{r=a} = 0. \tag{6.16}$$

The scattering cross-section involves the derivatives of Bessel functions, and in the long wave limit $ka \to 0$ the first three partial waves $n = 0 \pm 1$ all contribute to the leading term in $\sigma$ and give

$$\sigma \sim \frac{3\pi^2}{4} (ka)^3 a. \tag{6.17}$$

This method of calculation can be generalized to deal with the scattering by any body for which there exists an orthogonal coordinate system such that the body coincides with one of the surfaces $u_i = $ constant, though complications may occur because in three-dimensional problems the vector wave equation for the fields, rather than the scalar equation, must be used. In the problem of a cylinder we have just considered, the vector equation reduced to a scalar one because the problem was effectively only two dimensional. The procedure amounts to finding product solutions which can be used to satisfy the boundary conditions, and the result will usually be in the form of an infinite series. In particular the problem of the scattering of a plane wave by a conducting sphere can be solved in this way. The scattered wave is a linear combination of spherical waves of the form (2.58) (2.59) in which both T.E. and T.M. waves must be included. The amplitudes of the partial waves are determined by expressing the plane wave as a linear combination of spherical waves (this step corresponds to (6.7)), and then by using the boundary condition that the total tangential electric field must vanish on the surface of the sphere $r = a$.

In the long wave limit $(ka \ll 1)$ the total scattering cross-section for the sphere behaves as

$$\sigma \sim \frac{10\pi}{3} k^4 a^6, \tag{6.18}$$

so the scattering is inversely proportional to the fourth power of the wavelength $\lambda$ of the incident wave. This same dependence on $\lambda$ occurs for the scattering by a dielectric sphere and (6.18) can be used as a basis for an explanation for the colour of the sky. The short wavelength blue waves of sunlight are scattered very much more than the longer red waves by the air molecules and extremely small water droplets in the atmosphere. At sunset, when the ray path through the atmosphere is at its greatest, the blue part of the spectrum may be almost completely scattered out of the direct rays so that the sun appears an intense red.

## 6.3 Fresnel and Fraunhofer diffraction

Several problems involving the scattering of a wave by an edge have been solved exactly. The first exact solution was given by Sommerfeld for the scattering of a plane wave by the straight edge of an infinitesimally thin, semi-infinite plane screen (see [6]). However, the mathematical techniques required for the solution are rather beyond the level required for the rest of this text, so we shall not present the solution here. We shall comment on some of the difficulties presented by this problem later. Instead, we shall consider an approximate method, which comes under type $(c)$ in our classification of approximations given in the introduction, and use it to consider the diffraction of a plane wave by an aperture of arbitrary shape in an infinite plane screen. The method is based on Helmholtz's solution of the wave equation (4.24).

The Hertz potential $Z$ corresponding to a wave propagated within a homogeneous dielectric medium ($\varepsilon$, $\mu$) must satisfy a vector wave equation. If there are no sources, and if the wave has a harmonic time dependence $e^{i\omega t}$, then $Z$ satisfies the homogeneous equation,

$$\nabla^2 Z + \mu\varepsilon\omega^2 Z = 0. \tag{6.19}$$

Each cartesian component of $Z$ satisfies the corresponding scalar wave equation and so we can use the method of solution given in (4.2) to express $Z$ in terms of $Z$ and $\dfrac{\partial Z}{\partial n}$ over any closed surface $S$.

If the observation point $r$ is situated outside the surface $S$, and if each component of $Z$ satisfies the radiation condition (4.29), then the appropriate solution for $Z(r, t)$ is

$$Z(r, t) = \frac{1}{4\pi} \int_s \left\{ Z(r', t)\frac{\partial}{\partial n}\left(\frac{e^{-ikR}}{R}\right) - \frac{e^{-ikR}}{R}\frac{\partial Z(r', t)}{\partial n} \right\} dS', \tag{6.20}$$

where $R = |\,r - r'\,|$. This equation can be obtained from an equation corresponding to (4.24) by multiplying by $e^{i\omega t}$, together with a change of sign for the surface integral because in this case we require the solution in the exterior region. The inhomogeneous term is absent because there are no sources outside $S$.

A plane harmonic wave of frequency $\omega$ travelling from the negative $z$-direction can be described by a Hertz potential $Z_0$ given by

$$Z_0(r, t) = E_0 e^{i(\omega t - kz)}, \tag{6.21}$$

where $E_0$ is a constant vector in the $y$-direction. This form for $Z_0$ satisfies (6.19) and has associated electric and magnetic fields given by

$$E = \frac{\omega^2}{v^2}\, Z_0, \qquad H = \frac{\omega^2}{\mu v^3}\, k_0 \wedge Z_0, \tag{6.22}$$

where $k_0$ is a unit vector in the $z$-direction and $v^2 = 1/\mu\varepsilon$. If this wave falls normally on to an infinite plane screen $S_c$ (in $z = 0$) containing an aperture $S_g$, there will be some form of wave propagated into the region $z > 0$. We can relate the Hertz potential of the wave in the region $z > 0$ to the values of $Z$ and $\dfrac{\partial Z}{\partial n}$ over the exterior surface of the screen $S_c$, and the gap $S_g$, by using (6.20). If it were possible to

have a completely absorbing screen such that on the exterior surface
of $S_c$,

$$Z = 0, \qquad \frac{\partial Z}{\partial n} = 0, \tag{6.23}$$

we would be able to relate $Z(r, t)$ for $z > 0$ to the values in the
aperture alone

$$Z(r, t) = \frac{1}{4\pi} \int_{S_g} \left\{ Z'(r', t) \frac{\partial}{\partial n} \left( \frac{e^{-ikR}}{R} \right) - \frac{e^{-ikR}}{R} \frac{\partial Z(r', t)}{\partial n} \right\} dS'. \tag{6.24}$$

If further we could assume that the potential $Z(r', t)$ within the
aperture is given by $Z_0(r', t)$ the value that the potential would have
if the screen were absent, then (6.24) would give us the approximate
solution for $Z(r, t)$ in the region $z > 0$

$$Z(r, t) = \frac{1}{4\pi} \int_{S_g} \left\{ Z_0(r', t) \frac{\partial}{\partial n} \left( \frac{e^{-ikR}}{R} \right) - \frac{e^{-ikR}}{R} \frac{\partial Z(r', t)}{\partial n} \right\} dS'. \tag{6.25}$$

We can present one or two plausibility arguments in an attempt to
justify this approximation for waves of wavelengths small in comparison with the aperture size.

1. In the limit $k \to \infty$ (wavelength $\lambda \to 0$), geometrical optics can be
applied. In this limit, $S_c$ produces a complete shadow and the waves
in the aperture are unaffected by the presence of the screen.

2. The diffraction pattern deduced from (6.25) will not depend upon
the nature of the screen. This is consistent with experimental observa-
tions on apertures which are large in comparison with the wavelength.

To be set against these plausibility arguments are mathematical
objections to the assumption (6.23). It can be proved that, for solutions of the homogeneous wave equation valid with a region $\mathscr{R}$, this
condition cannot be imposed over a finite surface without implying
$Z = 0$ throughout $\mathscr{R}$. We omit a proof of this because it would
require a lengthy digression into mathematical analysis. This particular difficulty can be circumvented by introducing a suitably defined
Green's function. It involves a return to the derivation of (4.24) and
the use of a Green's function $G(R)$ instead of $e^{-ikR}/4\pi\varepsilon R$, for $\psi_2$ in

Green's Theorem (4.15). The Green's function is defined by

$$\nabla^2 G + k^2 G = 0 \qquad \text{outside } S, \tag{6.26}$$

$$G \to \frac{e^{-ikR}}{4\pi\varepsilon R} \qquad \text{as } R \to 0, \tag{6.27}$$

and

$$R\left(\frac{\partial G}{\partial n} + ikG\right) \to 0 \qquad \text{as } R \to \infty, \tag{6.28}$$

so that all the arguments are the same as in the previous derivation. The difference comes when we place a further boundary condition on G,

$$G = 0 \qquad \text{on } S \tag{6.29}$$

We denote the Green's function that satisfies this equation by $G_I$.
We could equally well have chosen $G$ to satisfy the condition,

$$\frac{\partial G}{\partial n} = 0 \qquad \text{on } S, \tag{6.30}$$

which defines an alternative Green's function $G_{II}$. If we rederive (6.21) using $G_I$, we find

$$Z(\mathbf{r}, t) = \varepsilon \int_S Z(\mathbf{r}', t)\frac{\partial G_I}{\partial n}\, dS', \tag{6.31}$$

and, if we use $G_{II}$, we obtain an alternative form

$$Z(\mathbf{r}, t) = -\varepsilon \int_S G_{II}\frac{\partial Z(\mathbf{r}', t)}{\partial n}\, dS'. \tag{6.32}$$

Both these forms enable us to avoid the difficulties involved in the approximation (6.23). If we assume that the screen is conducting and earthed, and that $Z$ on the screen is independent of the $x$ and $y$ coordinates, we may take $Z = 0$, so that

$$Z(\mathbf{r}, t) = \varepsilon \int_{S_g} Z(\mathbf{r}', t)\frac{\partial G_I}{\partial n}\, dS'. \tag{6.33}$$

The calculation of $G_I$ is not a trivial matter. Fortunately, when the condition (6.29) is imposed on a plane surface $G_I$ can be obtained by the method of images, a technique the reader must have met before in applications to simple electrostatic problems. If a unit source of the wave equation (6.26) is located at the point $(x, y, z)$ and a negative image source is located on the opposite side of the screen at the point $(x, y, -z)$, then $G_I$ at any point $(x', y', z')$ is given by

$$G_I = \frac{1}{4\pi\varepsilon}\left\{\frac{e^{-ikR}}{R} - \frac{e^{-ikR_1}}{R_1}\right\}, \tag{6.34}$$

where $R_1^2 = (x-x')^2+(y-y')^2+(z+z')^2$. The contributions from the two sources cancels on the plane $z' = 0$ so $G_I$ satisfies (6.29). The singularity at $R_1$ does not conflict with conditions (6.26) (6.27) (6.28) because it is located in the 'interior' region of $S$.

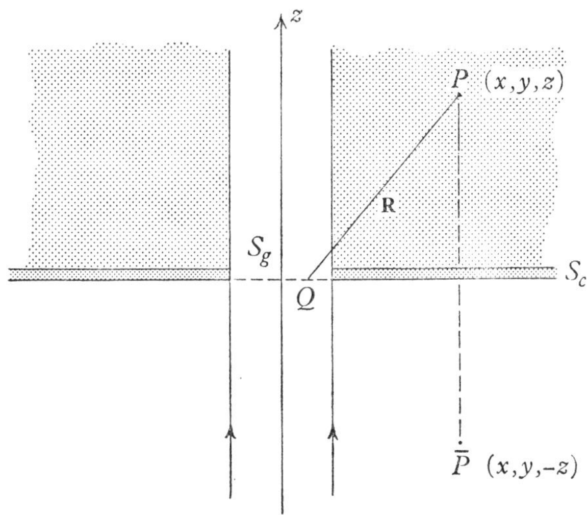


Fig. 19 The point $P$ is the observation point. $\bar{P}$ is its image point in the screen $S_c$. $Q$ is a point $(x', y', 0)$ in the aperture $S_g$

The normal derivative of $G_I$ on the plane $z' = 0$ is given by

$$\frac{\partial G_I}{\partial n} = -\frac{n \cdot R}{2\pi\varepsilon R^2}\left(ik + \frac{1}{R}\right)e^{-ikR}. \tag{6.35}$$

For waves of very short wavelength $(kR > 1)$ we need only retain the leading term of (6.35). On making the substitution into (6.33) we find

$$Z(r, t) = -\frac{ik}{2\pi}\int_{S_g} Z(r', t)\frac{n \cdot R}{R^2}e^{-ikR}\,dS'. \tag{6.36}$$

With the approximation that $Z(r', t)$ within the aperture is given by

95

the Hertz potential of the incident wave (6.21), we obtain an approximate expression for the potential associated with the diffraction fields, in the region $z>0$, given by

$$Z(r, t) \simeq - \frac{ikE_0 e^{i\omega t}}{2\pi} \int_{S_g} \frac{\boldsymbol{n} \cdot \boldsymbol{R}}{R^2} e^{-ikR} dS'. \tag{6.37}$$

As we move the point $(x', y', 0)$ across the aperture the exponential term in the integrand will vary very rapidly because the aperture size, it has been assumed, is very much larger than the wavelength. In comparison the term $\boldsymbol{n} \cdot \boldsymbol{R}/R^2$ will vary much more slowly and to a good approximation may be taken outside the integral so that

$$Z(r, t) \simeq - \frac{ikE_0 \boldsymbol{n} \cdot \boldsymbol{R} e^{i\omega t}}{2\pi R^2} \int_{S_g} e^{-ikR} dS'. \tag{6.38}$$

If the $z$-axis passes through the aperture and the distance of the point of observation $r$ is such that $x', y' \ll r$, we can expand $R$ in powers of $x'/r$ and $y'/r$ as

$$R \sim r - \frac{(xx'+yy')}{r} + \frac{(x'^2+y'^2)}{2r} - \frac{(xx'+yy')^2}{2r^3} + .... \tag{6.39}$$

For observation points at some distance from the $z$-axis $(x \gg x', y \gg y')$ we can neglect the quadratic terms in (6.39), and write

$$R \sim r - (lx' + my') + ..., \tag{6.40}$$

where $l$ and $m$ are direction cosines from the origin to the observation point $r$. If we substitute (6.40) into (6.38), the Hertz potential we obtain will be proportional to the integral

$$\int_{S_g} e^{ik(lx' + my')} dx' dy'. \tag{6.41}$$

The diffraction pattern in the region where the approximation (6.40) is valid is known as Fraunhofer diffraction. To determine the absolute intensity of the diffracted waves, calculation of the fields from the Hertz potential is necessary. These fields will be proportional to (6.41) so the *relative intensity* of the diffraction pattern will largely be determined by the square modulus of the integral (6.41).

For a rectangular aperture of dimensions $2a$ and $2b$ (in the $x'$ and $y'$ directions respectively), centred at the origin, the integral can be evaluated analytically

$$\int_{-a}^{a} e^{iklx'}dx' \int_{-b}^{b} e^{ikmy'}dy' = \frac{4\sin(kla)}{kla} \times \frac{\sin(kmb)}{kmb}. \qquad (6.42)$$

The relative intensity of the diffraction pattern will be determined by

$$\frac{\sin^2(kla)\sin^2(kmb)}{(kla)^2(kmb)^2}. \qquad (6.43)$$

The pattern on a large screen placed parallel to the diffracting screen will consist of spots of varying intensities forming a rectangular grid. The spots will be very approximately rectangular in shape, and the size of the pattern will be proportional to the distance of the observation screen from the aperture, because (6.43) depends only on the direction cosines. The zeros in intensity occur at points at which either $kla$ or $kmb$ is equal to an integer multiplied by $\pi$. The central spot will be by far the brightest and, as we move from the centre of the screen parallel to the $x$ (or $y$) axis, the maximum intensity of the second, third and fourth bright spots occur at values of $kla$ (or $kmb$) equal to $1{\cdot}43\pi$, $2{\cdot}46\pi$, and $3{\cdot}47\pi$, with intensities relative to the central spot of $0{\cdot}047$, $0{\cdot}017$ and $0{\cdot}008$ respectively. The intensities of spots which do not lie on the $x$ or $y$ axes fall off even more rapidly.

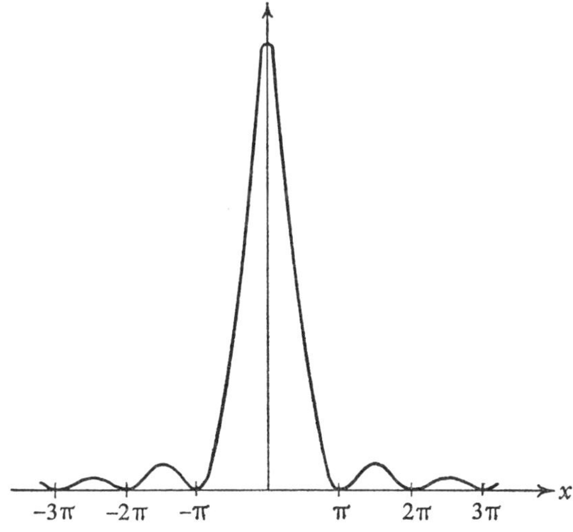


Fig. 20 The function $\left(\dfrac{\sin x}{x}\right)^2$ that determines the form of the Fraunhofer diffraction pattern from a rectangular aperture

If the aperture is circular, centred on the $z$-axis, with radius $a$ the diffraction pattern will be determined by the integral,

$$\int_0^a \int_0^{2\pi} e^{ik\rho s \cos (\theta' - \theta)} \rho\, d\theta'\, d\rho \qquad (6.44)$$

where for convenience we have introduced the polar coordinates $x' = \rho \cos \theta'$, $y' = \rho \sin \theta'$, $x = rs \cos \theta$, and $y = rs \sin \theta$. The $\theta'$-integral corresponds to the integral representation of the Bessel function $J_0(k\rho s)$ of zero order, hence (6.44) can be written in the form,

$$2\pi \int_0^a J_0(k\rho s)\rho\, d\rho. \qquad (6.45)$$

By using the recurrence relation,

$$\frac{d}{dx}(x^{n+1}J_{n+1}(x)) = x^{n+1}J_n(x), \qquad (6.46)$$

for Bessel functions (see Appendix B), (6.45) can be integrated and the square modulus of the result is

$$4\pi a^4 \left\{ \frac{J_1(ksa)}{ksa} \right\}^2. \qquad (6.47)$$

The behaviour of this function is not unlike the function $\left( \dfrac{\sin x}{x} \right)^2$.

There is no dependence on $\theta$ so the diffraction pattern consists of a series of alternate light and dark concentric circular fringes. The central circular spot accounts for most of the total intensity. The first dark ring that surrounds it has zero intensity at $ksa = 1{\cdot}22\pi$. The maximum intensity of the second bright ring occurs for $ksa = 1{\cdot}64\pi$, but is only $0{\cdot}018$ relative to the intensity of the central spot. The relative intensity drops to $0{\cdot}004$ at the third bright ring.

Diffraction effects can limit the resolving power of optical instruments. For example, if light of wavelength $\lambda(= 2\pi/k)$ passes through the circular objective lens of a telescope of radius $a$, the radial angular width of the central spot in the diffraction pattern will be $0{\cdot}61\lambda/a$. The angular separation of two point stars viewed with the microscope must exceed $0{\cdot}61\lambda/a$ if the observed diffraction pattern is to be resolved into two separate images. The value $0{\cdot}61\lambda/a$ will be the limiting angular resolution of the instrument.

For observation points near the $z$-axis or points much closer to the diffracting screen the quadratic terms in (6.39) cannot be neglected. Returning to the general diffraction integral (6.38) we observe that the rapid variations of $e^{-ikR}$ tend to cancel out in small regions about most points in the aperture. At the special point $(x', y')$ for which $R(x', y')$ is a stationary function, the exponential varies less rapidly and the main contribution to the diffraction integrals comes from the region around this point. When the point of observation $P$ lies in the central beam (see Fig. 19) $R$ is obviously stationary at the point $x' = x$, $y' = y$ when $R$ is parallel to the $z$-axis. If we expand $R$ in powers of $x'-x$ and $y'-y$ about this particular point,

$$R \sim z - \frac{1}{2z} \{(x'-x)^2 + (y'-y)^2\} + ..., \tag{6.48}$$

the linear terms are absent. Substituting (6.48) into (6.38) we find the integral that determines the form of the diffraction is

$$\int_{-a}^{a} e^{\frac{ik}{2z}(x'-x)^2} dx' \int_{-b}^{b} e^{\frac{ik}{2z}(y'-y)^2} dy'. \tag{6.49}$$

The diffraction pattern in the region where (6.49) is a valid approximation is known as Fresnel diffraction. The two integrals in (6.49) have the form $C(u_1, u_2) + iS(u_1, u_2)$, where

$$C(u_1, u_2) = \int_{-u_2}^{u_1} \cos\left(\frac{\pi}{2} u^2\right) du, \quad S(u, u_2) = \int_{-u_2}^{u_1} \sin\left(\frac{\pi}{2} u^2\right) du, \tag{6.50}$$

and are known as Fresnel's integrals. To transform the first integral in (6.49) to this form, we make the substitution $u = \sqrt{\frac{k}{z\pi}}(x'-x)$,

$u_1 = \sqrt{\frac{k}{z\pi}}(a-x)$, and $u_2 = \sqrt{\frac{k}{z\pi}}(a+x)$. These integrals have been tabulated because they cannot be evaluated analytically. The qualitative behaviour of $C(u_1, u_2)$ and $S(u_1, u_2)$ as functions of $u_1$ and $u_2$ can be seen quite easily. The integrand of each integral oscillates with constant amplitude as $u$ is varied. However, the distance between consecutive oscillations decreases as $u$ increases, and tends to zero as $2/u$ in the limit $u \to \infty$. Consequently, if $u_2 = 0$ and $u_1$ increases from zero, the integrals increase slowly and begin to oscillate. As $u_1$ gets larger the oscillations become more frequent but less pronounced, and die away as $u_1 \to \infty$. The integral in this limit exists because the contributions from the region of large $u$ tend to cancel out. By substituting $\beta = i$ in the well-known result

$$\int_0^\infty e^{-\beta x^2}\, dx = \tfrac{1}{2}\sqrt{\frac{\pi}{\beta}} \qquad (6.51)$$

we find from the real and imaginary parts that $C(\infty, 0) = \tfrac{1}{2}$, $S(\infty, 0) = \tfrac{1}{2}$.

For points $(x, y, z)$ lying within the central beam $u_1$ and $u_2$ have the same sign. As we move from the edge of the beam towards the centre the intensity oscillates as illustrated in Fig. 20. If we move from the edge into the region of shadow $u_1$ and $u_2$ have opposite signs and the square modulus of (6.49) slowly decreases to zero. It is clear from Fig. 19 that in this region $R$ has no stationary values with respect to $x'$ and $y'$ that lie within the aperture.

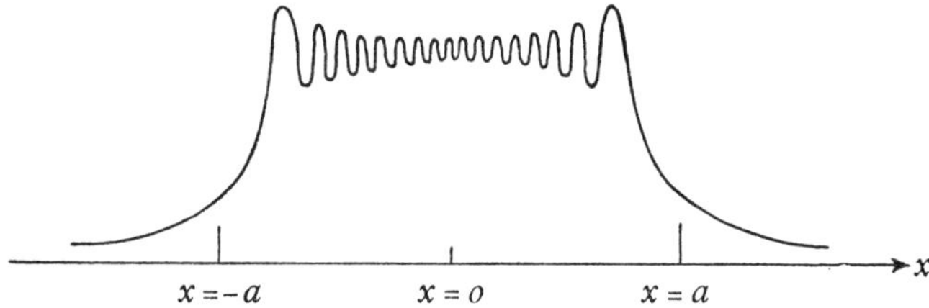


Fig. 21 The intensity of the Fresnel diffraction pattern across a rectangular aperture of width $2a$

As the observation screen is moved away from the aperture the diffraction pattern extends further into the region of shadow and changes continuously into the Fraunhofer form.

Calculations based on Helmholtz's solution seem to be entirely adequate for most optical problems, provided the circumstances comply with the assumptions made in the derivation, and the method is widely used. It is not entirely satisfactory from a mathematical point of view, however, because the calculation of the fields at the screen do not reproduce the assumed values except in the limit $\lambda \to 0$ $(k \to \infty)$. This suggests that the method might be a first approximation in an iteration scheme for small $\lambda$, but investigations show that this is not the case (see [7]).

A solution to a diffraction problem must satisfy the boundary conditions on the screen if it is to be a satisfactory solution from a mathematical point of view. To keep these boundary conditions reasonably simple an idealized perfectly conducting, and hence perfectly reflecting screen is usually considered. It is also assumed to be infinitesimally thin, an assumption which can be a source of difficulty because further conditions must be imposed to ensure that the mathematical problem has a unique solution in this limit. Several idealized problems have been solved exactly. They are important from a theoretical point of view and for the insight they give into the limitations of approximate techniques which are more generally applicable.

## PROBLEMS 6

1. Use the method described in section 6.2 to examine the scattering of an harmonic plane wave of frequency $\omega$ incident on an infinite dielectric cylinder $(\varepsilon_1, \mu_1)$ of radius $a$ in a direction perpendicular to its axis. Assume the plane wave is polarized such that its electric vector is parallel to the axis of the cylinder.

2. A diffraction grating is composed of $N$ rectangular slits, each of width $a$ along the $x$-axis and of length $b$ along the $y$-axis. If the slits are equally spaced with their centre along the $x$-axis at distances $d$ apart, show that the intensity of the diffraction pattern at a point in the Fraunhofer region with direction cosines $(l, m)$ is equal to that of a single slit multiplied by a factor

$$\frac{\sin^2 (\pi l N d/\lambda)}{\sin^2(\pi l d/\lambda)},$$

for a plane wave of wavelength $\lambda$ incident from the $z$-direction.

3. Show for a point $(x, y)$ on a curve which has the parametric equation

$$x(u) = \int_0^u \cos\left(\frac{\pi}{2}u^2\right) du, \qquad y(u) = \int_0^u \sin\left(\frac{\pi}{2}u^2\right) du,$$

that $u$ is the length of the curve from $(x, y)$ to the origin, and that $\frac{dy}{dx} = \tan\left(\frac{\pi}{2}u^2\right)$. Hence, show that the curve has the form of a spiral about the two points $(\frac{1}{2}, \frac{1}{2})$ and $(-\frac{1}{2}, -\frac{1}{2})$. This curve is known as 'Cornu's Spiral'. The distance between two points $u_1$ and $u_2$ on the spiral is equal to the modulus of $C(u_1, u_2) + iS(u_1, u_2)$, so the curve can be used to examine the form of the Fresnel diffraction pattern (6.49).

# APPENDIX A

# Units and Fundamental Constants

The first two systems of units devised for the measurement of electrical quantities were described as 'absolute', because all the quantities could be expressed solely in terms of units of mass ($M$), length ($L$), and time ($t$). The two systems are the *electrostatic* system (e.s.u.) and the *electromagnetic* system (e.m.u.); both are based on the three units, the centimetre, the gram and the second. The electrostatic system takes Coulomb's law

$$F_{12} = \frac{1}{4\pi\varepsilon_0} q_1 q_2 \frac{r_1 - r_2}{|r_1 - r_2|^3}, \tag{A.1}$$

as its starting-point. The constant of proportionality $1/4\pi\varepsilon_0$ is taken to be dimensionless and of unit magnitude so that from this equation charge can be expressed in terms of force and length, and ultimately in centimetres, grams and seconds. Once the unit of charge is defined, units for other electrical quantities can be deduced, including the constant of proportionality $\mu_0/4\pi$ in equation (1.22)

$$B = \frac{\mu_0}{4\pi} \int_v \frac{J(r') \wedge (r - r')}{|r - r'|^3} \, d\tau'. \tag{A.2}$$

The dimension of all the other quantities in equation (A.2) are defined and, in consequence, $\mu_0$ has the dimensions $(t/L)^2$. The value of $\mu_0$ cannot be chosen arbitrarily for the same reason, and has to be determined by experiment.

In the electromagnetic system, on the other hand, $\mu_0/4\pi$ is taken to be dimensionless and equal to unity. This enables the unit of current to be defined because (A.2) can be used to express the force $F_{12}$ between two current distributions, $J_1$ and $J_2$ as

$$F_{12} = \frac{\mu_0}{4\pi} \iint \frac{J_2 \wedge (J_1 \wedge (r_1 - r_2)) d\tau_1 d\tau_2}{|r_1 - r_2|^3}. \tag{A.3}$$

The electromagnetic units of all the other quantities can all be calculated from this starting-point. In e.m. units $1/4\pi\varepsilon_0$ has the dimensions $(t/L)^2$ and must be experimentally determined.

In these two systems there is essentially only one constant $\varepsilon_0\mu_0$ that has to be found experimentally. In both systems $(\mu_0\varepsilon_0)^{-\frac{1}{2}}$ has the dimensions of velocity $(L/t)$, and its measured value is $2\cdot998\times10^{10}$ cm/sec, the velocity $(c)$ of light in free space. It was the measurement of this quantity which was so significant and led to Maxwell's conjecture that the wave equations (2.8) and (2.9) described light waves.

Other units have been derived from these two absolute systems. One is the Gaussian system which involves mixed units; $B$ and $H$ are measured in e.m. units while all other quantities are measured in e.s.u. The constants $\varepsilon_0 = 1$ and $\mu_0 = 1$ in this system, and the factor $c$ is included in some of the equations as a conversion factor from e.s.u. to e.m.u. Until quite recently this was a popular system for mathematical texts on electromagnetism.

For the convenience of expressing the measured values of electrical quantities a system of *practical* units was devised. These include the Coulomb (for charge), the volt (potential) and the ampere (current), and all the units are related to the corresponding quantities in e.m.u. by factors of ten.

In recent years there has been a concerted effort to replace these four rival systems by an internationally agreed system, S.I. units, which combine many of the better features of the previous systems.

The forerunner of S.I. units was the M.K.S. system. This system was based on a suggestion by Giorgi in 1901, who showed that an 'absolute' system could be devised which could result in units of a magnitude convenient for many practical purposes if some electrical unit could be taken as a fourth standard unit to be combined with the metre, kilogram, and second, the units of length, mass and time. This idea was officially adopted in 1950 by the International Electro-technical Commission, when the ampere $(A)$ was adopted as the standard electrical quantity. In 1954, the Conférènce Générale des Poids et Mesures (C.G.P.M.), which is responsible for international matters concerning the metric system, adopted the six standard units, the metre, kilogram, second, ampere, the degree Kelvin as a unit of temperature, and the candela as the unit of luminous intensity. In 1960, the C.G.P.M. gave the system the full title 'Système International d'Unités' (S.I.) and since that time many scientific journals have made it their policy only to accept papers using these units except where special considerations apply.

The S.I. unit system is rationalized, i.e. factors of $4\pi$ are introduced with $\varepsilon_0$ and $\mu_0$ so as to simplify many of the equations; in particular, the factors of $1/4\pi$ which appear in Maxwell's equations (1.3) and (1.4) in unrationalized units are eliminated.

The dimensions of all electrical quantities can be expressed in terms of the four dimensions, $M$, $L$, $t$ and $A$; which we show in the table below, together with the S.I. description of the unit.

| *Quantity* | *Dimensions* | *S.I. unit* |
|---|---|---|
| *Basic units* | | |
| Length | $L$ | Metre |
| Mass | $M$ | Kilogram |
| Time | $t$ | Second |
| Electric current | $A$ | Ampère |
| Absolute temperature | $K$ | °Kelvin |
| Luminous intensity | $C$ | Candela |
| *Derived units* | | |
| Force ($F$) | $\dfrac{ML}{t^2}$ | Newton |
| Energy ($U$) | $\dfrac{ML^2}{t^2}$ | Joule |
| Power | $\dfrac{ML^2}{t^3}$ | Watt |
| Charge ($q$) | $At$ | Coulomb |
| Electric Potential ($\phi$) | $\dfrac{ML^2}{At^3}$ | Volt |
| Electric field ($E$) | $\dfrac{ML}{At^3}$ | Volts/metre |
| Resistance | $\dfrac{ML^2}{A^2t^3}$ | Ohm |
| Conductivity ($\sigma$) | $\dfrac{A^2t^3}{ML^3}$ | Siemens/metre |
| Electric Polarization ($P$) | $\dfrac{At}{L^2}$ | Coulomb/metre$^2$ |
| Capacitance | $\dfrac{A^2t^4}{AL^2}$ | Farad |

# Appendices

| Quantity | Dimensions | S.I. unit |
|---|---|---|
| Dielectric constant ($\varepsilon$) | $\dfrac{A^2 t^4}{ML^3}$ | Farads/metre |
| Magnetic induction ($\boldsymbol{B}$) | $\dfrac{M}{ALt}$ | Weber |
| Magnetic field ($\boldsymbol{H}$) | $\dfrac{A}{L}$ | Ampères/metre |
| Inductance | $\dfrac{ML^2}{A^2 t^2}$ | Henry |
| Permeability ($\mu$) | $\dfrac{ML}{A^2 t^2}$ | Henries/metre |

For reference, we give a list of the values of relevant fundamental constants.

$$\text{Charge on an electron} = 1{\cdot}6 \times 10^{-19} \text{ coulombs}$$

$$\text{Mass of an electron} = 9{\cdot}1 \times 10^{-31} \text{ kilograms}$$

$$\text{Mass of a proton} = 1{\cdot}67 \times 10^{-27} \text{ kilograms}$$

$$\mu_0 = 4\pi \times 10^{-7} \text{ Henries/metre}$$

$$\varepsilon_0 = 8{\cdot}854 \times 10^{-12} \text{ farads/metre}$$

$$c = 2{\cdot}997 \times 10^8 \text{ metres/second.}$$

$$\textbf{APPENDIX B}$$

# Mathematical Compendium

For reference purposes we give a brief list of some of the theorems and formulae used in the text.

**1. Vector operators:** The formulae given below apply at points within a domain $\mathcal{R}$ in which the scalar function $\phi(x, y, z)$ and $\psi(x, y, z)$, and the vector functions $A(x, y, z)$ and $B(x, y, z)$, are defined and are continuous with continuous first and second order partial derivatives. These conditions are sufficient but not in all cases necessary. For a more precise statement concerning their validity see [8] in the bibliography.

In cartesian coordinates grad $\phi$ is defined by

$$\text{grad } \phi = i\,\frac{\partial \phi}{\partial x} + j\,\frac{\partial \phi}{\partial y} + k\,\frac{\partial \phi}{\partial z}, \tag{B.1}$$

where $i, j$ and $k$ are unit vectors in the $x$, $y$ and $z$ directions respectively. The divergence of a vector $A$ is defined as

$$\text{div } A = \frac{\partial A_1}{\partial x} + \frac{\partial A_2}{\partial y} + \frac{\partial A_3}{\partial z}, \tag{B.2}$$

and curl $A$ by

$$\text{curl } A = i\left(\frac{\partial A_3}{\partial y} - \frac{\partial A_2}{\partial z}\right) + j\left(\frac{\partial A_1}{\partial z} - \frac{\partial A_3}{\partial x}\right) + k\left(\frac{\partial A_2}{\partial x} - \frac{\partial A_1}{\partial y}\right), \tag{B.3}$$

where $A_1$, $A_2$ and $A_3$ are the cartesian components of $A$.

In any system of curvilinear coordinates $(u, v, w)$ the differential element of length $ds$ between the points $(x, y, z)$ and

$$(x+dx, y+dy, z+dz)$$

can be expressed in terms of the differentials $du$, $dv$, and $dw$ by an equation of the form,

$$ds^2 = h_1^2 du^2 + h_2^2 dv^2 + h_3^2 dw^2 + 2h_{12}dudv + 2h_{23}dvdw + 2h_{13}dudw, \tag{B.4}$$

where $h_1$ etc. are functions of $u$, $v$ and $w$. In an orthogonal system $h_{12} = h_{23} = h_{13} = 0$, and in this coordinate system grad $\phi$ is given by

$$\text{grad } \phi = \frac{i_1}{h_1} \frac{\partial \phi}{\partial u} + \frac{i_2}{h_2} \frac{\partial \phi}{\partial v} + \frac{i_3}{h_3} \frac{\partial \phi}{\partial w}, \tag{B.5}$$

where $i_1$, $i_2$ and $i_3$ are unit vectors through the point $(u, v, w)$ in the directions $u = \text{constant}$, $v = \text{constant}$ and $w = \text{constant}$ respectively. Also we have

$$\text{div } A = \frac{1}{h_1 h_2 h_3} \left\{ \frac{\partial}{\partial u}(h_2 h_3 A_1) + \frac{\partial}{\partial v}(h_1 h_3 A_2) + \frac{\partial}{\partial w}(h_1 h_2 A_3) \right\}, \tag{B.6}$$

and curl $A$ expressed in determinant form

$$\text{curl } A = \frac{1}{h_1 h_2 h_3} \begin{vmatrix} h_1 i_1 & h_2 i_2 & h_3 i_3 \\ \dfrac{\partial}{\partial u} & \dfrac{\partial}{\partial v} & \dfrac{\partial}{\partial w} \\ h_1 A_1 & h_2 A_2 & h_3 A_3 \end{vmatrix}, \tag{B.7}$$

where $A_1$, $A_2$ and $A_3$ are components of $A$ along the directions $i_1$, $i_2$ and $i_3$.

For cylindrical coordinates $x = r \cos \theta$, $y = r \sin \theta$, "$z$" $= z$,

$$ds^2 = dr^2 + r^2 d\theta^2 + dz^2, \tag{B.8}$$

and hence $h_1 = 1$, $h_2 = r$ and $h_3 = 1$.

In the case of spherical polar coordinates $x = r \cos \psi \sin \theta$, $y = r \sin \psi \sin \theta$, $z - r \cos \theta$,

$$ds^2 = dr^2 + r^2 . d\theta^2 + r^2 \sin^2 \theta d\psi^2. \tag{B.9}$$

Therefore, $h_1 = 1$, $h_2 = r$ and $h_3 = r \sin \theta$.

The Laplacian operator $\nabla^2$ is defined as div grad and in cartesian coordinates has the form

$$\nabla^2 \phi = \frac{\partial^2 \phi}{\partial x^2} + \frac{\partial^2 \phi}{\partial y^2} + \frac{\partial^2 \phi}{\partial z^2}, \tag{B.10}$$

and for orthogonal curvilinear coordinates

$$\nabla^2 \phi = \frac{1}{h_1 h_2 h_3} \left\{ \frac{\partial}{\partial u}\left(\frac{h_2 h_3}{h_1} \frac{\partial \phi}{\partial u}\right) + \frac{\partial}{\partial v}\left(\frac{h_3 h_1}{h_2} \frac{\partial \phi}{\partial v}\right) + \frac{\partial}{\partial w}\left(\frac{h_1 h_2}{h_3} \frac{\partial \phi}{\partial w}\right) \right\}. \tag{B.11}$$

The operator $\nabla^2$, which operates on vector quantities, is defined by

$$\nabla^2 A = \text{grad div } A - \text{curl curl } A \qquad \text{(B.12)}$$

In the special case of cartesian coordinates,

$$\nabla^2 A = i\nabla^2 A_1 + j\nabla^2 A_2 + k\nabla^2 A_3.$$

**2. The divergence theorem:** The integral of the outward normal component of the vector $A$ over a closed surface $S$, limiting a volume $V$ is equal to the integral of div $A$ throughout $V$:—

$$\int_S A \cdot dS = \int_V \text{div } A d\tau. \qquad \text{(B.13)}$$

**3. Stokes's theorem:** The line integral of $A$ along a closed contour $C$ is equal to the normal component of curl $A$ integrated over any surface $S$ limited by $C$:—

$$\int_C A \cdot dr = \int_S \text{curl } A \cdot dS. \qquad \text{(B.14)}$$

**4. Green's theorem:** For a closed surface $S$ limiting a volume $V$:—

$$\int_S (\phi \text{ grad } \psi - \psi \text{ grad } \phi) \cdot dS = \int_V (\phi\nabla^2\psi - \psi\nabla^2\phi)d\tau. \qquad \text{(B.15)}$$

**5. Helmholtz's theorem:** A vector $A$ is uniquely determined in a region $\mathscr{R}$ limited by a closed surface $S$ if div $A$ and curl $A$ are given throughout $\mathscr{R}$ and $A_n$, the normal component of $A$ is given on S.

**6. Identities:** The following identities have been required:—

($a$) div $(A \wedge B) = B \cdot \text{curl } A - A \cdot \text{curl } B.$

($b$) curl $(A \wedge B) = A \text{ div } B - B \text{ div } A + (B \cdot \text{grad})A - (A \cdot \text{grad})B.$

($c$) curl grad $\phi = 0$. The converse is also important. If curl $A = 0$ (in $\mathscr{R}$), then there exists a scalar function $\phi$ (in $\mathscr{R}$) such that $A = \text{grad } \phi$ (assuming the continuity conditions on $A$ given in 1).

($d$) div curl $A \equiv 0$. The converse states that if div $B = 0$ (in $\mathscr{R}$) then there exists a vector $A$ such $B = \text{curl } A$ (in $\mathscr{R}$).

**7. Associated Legendre functions:** $P_n^m(x)$ and $Q_n^m(x)$ are solutions of the equation

$$(1-x^2)\frac{d^2y}{dx^2} - 2x\frac{dy}{dx} + \left(n(n+1) - \frac{m^2}{1-x^2}\right)y = 0. \qquad \text{(B.16)}$$

When $n$ and $m$ are non-negative integers $P_n^m(x)$ is a polynomial of degree $n-m$ for $n \geq m$ and zero for $m > n$. The polynomials corresponding to $m = 0$ are known as Legendre polynomials and are written as $B_n(x)$. For further properties of these functions see [9] and [10]. The first three Legendre polynomials are

$$P_0(x) = 1, \quad P_1(x) = x, \quad P_2(x) = \tfrac{3}{2}(3x^2 - 1).$$

**8. Bessel functions:** The Bessel functions $J_v(x)$ and $Y_v(x)$ of order $v$ are independent solutions of the equation

$$\frac{d^2y}{dx^2} + \frac{1}{x}\frac{dy}{dx} + \left(1 - \frac{v^2}{x^2}\right)y = 0. \tag{B.17}$$

The Bessel function $J_v(x)$ of the first kind can be written in the form of a series

$$J_v(x) = \sum_{r=0}^{\infty} \frac{(-1)^r \left(\dfrac{x}{2}\right)^{v+2r}}{(r+v)!\,r!}. \tag{B.18}$$

The function $Y_v(x)$ of the second kind is defined by

$$Y_v(x) = \lim_{\alpha \to v} \left\{ \frac{J_\alpha(x)\cos(\pi\alpha) - J_{-\alpha}(x)}{\sin(\pi\alpha)} \right\}. \tag{B.19}$$

Two useful recurrence relations for $J_v(x)$ are

$$xJ_v'(x) = xJ_{v-1}(x) - vJ_v(x),$$

and $\tag{B.20}$

$$xJ_v'(x) = vJ_v(x) - xJ_{v+1}(x).$$

When $n$ is an integer $J_n(x)$ can be represented by the integral form

$$J_n(x) = \frac{1}{2\pi} \int_{\alpha}^{2\pi + \alpha} e^{i(n\theta - x\sin\theta)} d\theta. \tag{B.21}$$

Spherical Bessel functions $j_l$ and $y_l$ are defined in terms of $J_{l+\frac{1}{2}}$ and $Y_{e+\frac{1}{2}}$ by

$$j_l(x) = \left(\frac{\pi}{2x}\right)^{\frac{1}{2}} J_{l+\frac{1}{2}}(x), \qquad y_l(x) = \left(\frac{\pi}{2x}\right)^{\frac{1}{2}} Y_{l+\frac{1}{2}}(x). \tag{B.22}$$

There are several very comprehensive accounts of the properties of Bessel functions (see [10] and also [11]).

**Fourier series:** A set of functions $\phi_n(x)$, where $n$ is an integer, defined over an interval $(a, b)$ are orthogonal if

$$\int_a^b \phi_n^*(x)\phi_m(x)dx \begin{cases} = 0, & n \neq m \\ = 1, & n = m. \end{cases} \tag{B.23}$$

A large class of functions $f(x)$ can be represented by a series of the form

$$f(x) = \sum_{n=-\infty}^{\infty} a_n \phi_n(x), \tag{B.24}$$

where

$$a_n = \int_a^b \phi_n f(x)dx, \tag{B.25}$$

if the set $\phi_n$ is orthogonal and complete (for a definition of *completeness*, sufficient conditions for the expansion, and a discussion of the sense in which the equality in (B.24) is to be interpreted see [10] and [12]).

The set of function $\dfrac{1}{\sqrt{2\pi}} e^{inx}$, where $n$ is any integer, form a complete orthogonal set over the interval $(0, 2\pi)$. The series for any function $f(x)$ defined over the interval $(0, 2\pi)$

$$f(x) = \sum_{n=-\infty}^{\infty} b_n e^{inx} \tag{B.26}$$

is the Fourier Series of $f(x)$, and coefficient $b_n$ is given by

$$b_n = \frac{1}{2\pi} \int_0^{2\pi} f(x)e^{-inx}\,dx. \tag{B.27}$$

The set of Legendre Polynomials $P_n(x)$ over the interval $(-1, 1)$ is another example of a complete orthogonal set.

**10. Fourier transforms:** The Fourier transform $f_\omega$ of a function $f(x)$ is defined by

$$f_\omega = \frac{1}{\sqrt{2\pi}} \int_{-\infty}^{\infty} f(x)e^{-i\omega x}\,dx, \tag{B.28}$$

provided the integral exists. The inverse transformation is

$$f(x) = \frac{1}{\sqrt{2\pi}} \int_{-\infty}^{\infty} f_\omega e^{i\omega x}\,d\omega. \tag{B.29}$$

By integrating by parts we can show that the Fourier transform of the derivative $\dfrac{df}{dx}$ is $i\omega f_\omega$. If $f(x)$ is an ordinary function $f_\omega \to 0$ as $\omega \to \pm\infty$. For the Fourier transform of generalized functions see [3].

# BIBLIOGRAPHY

References have been made in the course of the text to:

1. SOMMERFELD, A., *Ann. Physik*, **44**, 1914, 177.
   BRILLOUIN, L., *Ann. Physik*, **44**, 1914, 207.
2. GOOS, F. and HÄNCHEN, H., *Ann. Physik*, **6**, 1947, 1, 333.
3. LIGHTHILL, M. J., *Fourier Analysis and Generalized Functions*, Cambridge University Press 1958.
4. POWER, E. A., *Introductory Quantum Electrodynamics*, Longmans 1964.
5. TAMM, I. E. and FRANK, I. M., *C.R. Ac. Sci. U.S.S.R.* **14**, 1937, 109.
6. SOMMERFELD, A., *Math. Ann.* **47**, 1896, 317.
7. BOUWKAMP, C. J., *Rep. Prog. Phys.* **17**, 1954, 49.
8. RUTHERFORD, D. E., *Vector Methods*, Oliver and Boyd 1962.
9. WHITTAKER, E. T. and WATSON, G. N., *A Course of Modern Analysis*, Cambridge University Press 1965.
10. COURANT, R. and HILBERT, D., *Methods of Mathematical Physics*, Interscience 1963.
11. WATSON, G. N., *Theory of Bessel Functions*, Cambridge University Press 1922.
12. MESCHKOWSKI, H., *Series Expansions for Mathematical Physicists*, Oliver and Boyd 1968.


Detailed introductions to Maxwell's equations can be found in:

COULSON, C. A., *Electricity*, Oliver and Boyd 1958.

FEYNMAN, R. P., *Lectures on Physics*, Vols. I and II, Addison Wesley 1965.

PURCELL, E. M., *Electricity and Magnetism* (Berkeley Physics Course, Vol. 2), McGraw Hill 1965.


Some parallel and advanced texts for further reading:

BORN, M. and WOLF, E., *Principles of Optics*, Pergamon Press 1959.

BRILLOUIN, L., *Wave Propagation and Group Velocity*, Academic Press 1960.

JONES, D. S., *Theory of Electromagnetism*, Pergamon Press 1964.

# Bibliography

LANDAU, L. D. and LIFSHITZ, E. M., *The Classical Theory of Fields*, Pergamon Press 1961: *Electrodynamics of Continuous Media*, Pergamon Press 1960.

PANOFSKY, K. H. W. and PHILLIPS, M., *Classical Electricity and Magnetism*, Addison-Wesley 1956.

REITZ, J. R. and MILFORD, F. J., *Foundations of Electromagnetic Theory*, Addison-Wesley 1960.

SLATER, J. C. and FRANK, N. H., *Electromagnetism*, McGraw Hill 1947.

SOMMERFELD, A., *Optics*, Academic Press 1954.

STRATTON, J. A., *Electromagnetic Theory*, McGraw Hill 1941.

# INDEX